Green Chemistry in Environmental Sustainability and Chemical Education

Virinder S. Parmar · Priti Malhotra
Divya Mathur
Editors

Green Chemistry in Environmental Sustainability and Chemical Education

Proceedings of ICGC 2016, New Delhi

Springer

Editors
Virinder S. Parmar
Institute of Advanced Sciences
Dartmouth, MA
USA

Priti Malhotra
Department of Chemistry,
Daulat Ram College
University of Delhi
Delhi
India

Divya Mathur
Department of Chemistry,
Daulat Ram College
University of Delhi
Delhi
India

ISBN 978-981-13-4138-0 ISBN 978-981-10-8390-7 (eBook)
https://doi.org/10.1007/978-981-10-8390-7

Printed on acid-free paper

This Springer imprint is published by the registered company Springer Nature Singapore Pte Ltd.
The registered company address is: 152 Beach Road, #21-01/04 Gateway East, Singapore 189721, Singapore

Preface

Green chemistry is the inventions, design, and applications of chemical products and processes to reduce or to eliminate the use and generation of hazardous substances, maximize the conversion of the starting raw materials into the products, increase the energy efficiency, and minimize the waste production. Green chemistry is the need of the hour as it enables substantial progress toward equitable standards of living in a manner that is sustainable for future generations. It is therefore imperative to teach the value of green chemistry to future generations.

The International Conference on "Green Chemistry in Environmental Sustainability and Chemical Education (ICGC-2016)" held in Delhi on November 17 to 18, 2016, provided a platform for high school students, undergraduate and postgraduate students, teaching fraternity, and young researchers to discuss the role of chemistry in solving environmental challenges with a sustainable approach.

This volume of proceedings from the conference provides an opportunity for readers to engage with a selection of refereed papers that were presented during the conference. The 21 papers published here illustrate the multiple meanings of the terms sustainability and green chemistry. The reader will sample here reports of research on topics ranging from recent advances in green catalysis, environmental remediation, nanobiotechnology and innovative chemistry education, etc.

We hope that this proceeding volume will provide useful material and would be of immense use to educators, researchers, and students. We believe that the proceedings of ICGC-2016 would also play a vital role in teaching and inculcating responsible green chemistry practices in undergraduate and high school students who are our future researchers. We hope that the book provides enough interesting data, results, new achievements, and proactive approaches within green chemistry and environmental studies to any of readers' vocation and the field of interest.

Dartmouth, USA — Prof. Virinder S. Parmar
Delhi, India — Dr. Priti Malhotra
Delhi, India — Dr. Divya Mathur

Acknowledgements

The ICGC-2016 Organizing Committee would like to express immense gratitude to all the people and bodies that have helped us in the successful organization of the first ICGC conference. First of all, we would like to thank the Principal of Daulat Ram College, University of Delhi, Dr. Savita Roy, for her continuous support and advice which have greatly helped toward the successful organization of this conference. Our thanks go to all the governing body members and our Chairperson, Mrs. Suneeta Sudershan, for their constant encouragement. We are very much thankful to all the sponsors of this conference, Science and Engineering Research Board, Department of Science and Technology, Defence Research and Development Organization, and Department of Biotechnology, Government of India, for their generous financial support.

We would like to convey our sincere thanks to our Chief Guests, Dr. Chitra Rajagopal, Director, Centre for Fire, Explosive and Environment Safety, Ministry of Defence, Government of India, and Dr. Praveen Saxena, CEO, Skill Council for Green Jobs, for gracing the occasion. We also thank our Guest of Honor, Prof. Inder Mohan Kapahy, Honorable UGC Member, for his guidance and kind words.

We would like to express our sincere thanks and appreciation to the Conference Advisory Committee for its work and dedication. Special thanks also go to the keynote speakers and all the track chairs. A big thank-you goes to all the administrative staff that worked with us to ensure that we had a great event. We thank all oral presentation speakers and poster presenters for their enthusiastic participation in this conference. We acknowledge the unwavering support received from the faculty, staff members, and student volunteers. We appreciate and would like to thank Mr. Aninda Bose, Ms. Kamiya Khatter, Ms. Krati Shrivastav and their colleagues in Springer for their strong support towards publishing this proceedings. Our thanks also go to all the people who have given their precious time in making this conference a success.

About the Conference

Objective

The theme of the International Conference on "Green Chemistry in Environmental Sustainability and Chemical Education (ICGC-2016)" was to discuss the emerging green trends in the direction of sustainability and environmental safety. The United Nations General Assembly also addressed the environmental challenges in its Sustainable Development Goals (SDGs) which have been adopted in 2015. A closer look shows that to meet these goals, chemistry will play an important role.

ICGC-2016 was an initiative by the Daulat Ram College to create a platform for fruitful deliberations which will deeply enhance awareness about our responsibility toward the environment. The overarching theme was to discuss most recent innovations and concerns in green chemistry as well as practical challenges encountered and solutions adopted to remediate a scathed environment into a pristine one.

International Conference on "Green Chemistry in Environmental Sustainability and Chemical Education (ICGC-2016)" addressed the following topics:

- Green catalysis
- Green and sustainable processes
- Green environmental remediation
- Green materials
- Green nanotechnology
- Renewable energy
- Food process technology and engineering
- Industry and green chemistry for sustainability
- Eco-friendly green products
- Natural products
- Chemicals, chemistries, and processes to enable a circular economy

- Green analytical chemistry
- Biotechnology
- Environment and health
- Environmental management and policy
- Green chemistry education
- The UN Sustainable Development Goals and sustainable chemistry

Organizing Committee

Patron
Dr. Savita Roy, Principal, Daulat Ram College, University of Delhi

Convenor
Dr. Priti Malhotra, Department of Chemistry, Daulat Ram College

Co-convenor
Dr. Divya Mathur, Department of Chemistry, Daulat Ram College

Coordinator
Dr. Arti Jain, Department of Chemistry, Daulat Ram College

Associate Coordinator
Dr. Aditi Puri, Department of Chemistry, Daulat Ram College

Executive Committee Members
Dr. Shuchi Sharma
Dr. Archana Gupta
Dr. Asha
Dr. Darshan Singh
Dr. Ritu Gupta
Dr. Avneet Kaur
Ms. Sarita Singh
Dr. Amar Kumar
Dr. Preeti Chadha
Dr. Gauri Ahuja

Advisory Board Members

Keynote Summaries

Dr. Chitra Rajagopal

Outstanding Scientist and Director
Centre for Fire, Explosive and Environment Safety
DRDO, Ministry of Defence
New Delhi, India

Green Chemistry for Sustainable Development

Chitra Rajagopal*, Prasun Roy
Centre for Fire, Explosive and Environment Safety, DRDO
Ministry of Defence, New Delhi-54, India

It is the biggest challenge of the present generation to somehow satisfy society's needs without compromising the survival of future generations. In this endeavor to ensure that the development models we adopt are sustainable, green chemistry is increasingly seen as a crucial element that touches many aspects of the environment and human welfare. Defined as "**the utilisation of a set of principles that reduces or eliminates the use or generation of hazardous substances in the design, manufacture and application of chemical products**", its relevance to the 17 Sustainable Development Goals set by the United Nations last September is undisputable.

The phenomenal rise in human life expectancy and in the material quality of life that has come about in recent decades can be attributed to developments in chemistry and chemical engineering. However, these very developments have also led to several environmental and related disasters, some of them in our own country.

We now look at these disciplines to address the environmental problems of the past several decades. While environmental chemistry focuses on the effects of polluting chemicals on nature, green chemistry focuses on technological approaches to preventing pollution and reducing consumption of nonrenewable resources, such as

- Environmentally benign chemical synthesis and processes (green catalysis, green solvents and reagents, atom-economy synthetic methods, etc.)
- Green chemicals and energy produced from renewable resources (biomass, carbon dioxide, etc.)
- Novel materials and technologies for energy production and storage (biofuels, hydrogen, fuel cells, solar cells, lithium-ion batteries, etc.)
- Green chemical engineering processes (process integration, energy saving, waste minimization, efficient separation processes, etc.)
- Green technologies for environmental sustainability (carbon dioxide capture, HAZWASTE treatment, pollution prevention, environmental redemption, etc.)

CFEES, as the nodal laboratory for the implementation of environmental rules and regulations in MoD and for the implementation of the National Environment Policy in DRDO, has successfully delivered many products and technologies utilizing the principles of green chemistry and chemical engineering to address environmental problems, including clean energy generation from carbon neutral source, eco-friendly methods of treatment and disposal of hazardous wastes, recycling of PET wastes to generate MOFs, remediation of contaminated soils. CFEES is also looking to introduce the concept of sustainability through resource minimization and energy saving, in its developmental and infrastructural activities. CFEES plays a key role in capacity building through training programs on various aspects of safety management for MoD personnel. Recent initiatives include program on environmental safety and the introduction of sustainability concepts for DRDO personnel.

Prof. Christophe Len

Sorbonne Universités
Université de Technologie de Compiègne
France

Green Chemical Synthesis Around Furfural

Christophe Len*
Sorbonne Universités, Université de Technologie de Compiègne
Transformations Integrées de la Matière Renouvelable
Centre de Recherche de Royallieu, CS 60319, F-60203 Compiegne, France
Email: christophe.len@utc.fr

The design of environmentally friendly methodologies has been the driving force of scientists in recent years. In particular, the use of biomass-derived materials, green solvents, and alternatives techniques has been investigated.

In this conference, several green chemistry approaches that target advanced synthesis and processes will be presented. All these approaches include the production of furfural and derivatives from D-xylose, xylane, and hemicellulose using alternative technologies (e.g., microwave irradiation, high temperature/pressure) in batch and continuous flow via homogeneous and heterogeneous catalysis [1–4]. Conception, synthesis, and physicochemical properties will be detailed.

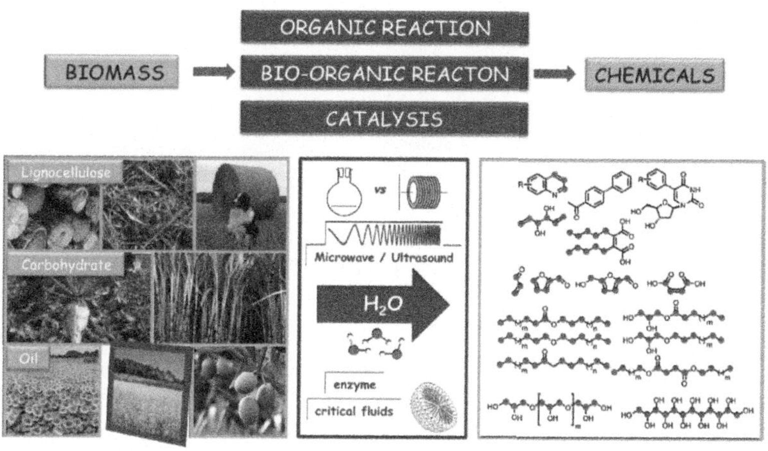

References

1. Le Guenic S, Delbecq F, Ceballos C, Len C* (2015) Microwave-aided dehydration of D-xylose into furfural by diluted inorganic salts solution in a biphasic system. J Mol Catal A: Chemical 410:1–7
2. Delbecq F, Wang Y, Len C (2016) Conversion of xylose, xylan and rice husk into furfural via betaine and formic acid mixture as novel homogeneous catalyst in biphasic system by microwave-assisted dehydration. J Mol Catal A: Chemical 423:520–525
3. Le Guenic S, Gergela D, Ceballos C, Delbecq F, Len C (2016) Furfural production from D-xylose and xylan by using stable Nafion NR50 and NaCl in a microwave-assisted biphasic reaction. Molecules 21:1102–1112
4. Wang Y, Len T, Huang Y, Taboada AD, Boa AN, Ceballos C, Delbecq C, Mackenzie, Len C (2016) Sulfonated Sporopollenin as an efficient and recyclable heterogeneous catalyst for dehydration of D-xylose and xylan into furfural. ACS Sustain Chem Eng (in press)

Prof. Virinder S. Parmar

Institute of Advanced Sciences
Dartmouth, USA

Greener and Sustainable Synthetic Routes to Novel Polymeric Materials, Nanocomposites, and Dendrons

Virinder S. Parmar*
Bioorganic Laboratory, Department of Chemistry
University of Delhi, New Delhi-110 007
India; Institute for Nanoscience and Nanomedicine,
University of Massachusetts, Lowell, MA 01854, USA;
Institute of Advanced Sciences, Dartmouth, MA 02747, USA
Email: virparmar@gmail.com; vparmar@inads.org

We have developed a chemoenzymatic synthesis for obtaining novel amphiphilic polymeric nanoparticles based on PEG having a broad range of additional chemical functionalities under mild conditions. Simplicity and versatility of this method for the synthesis of highly functionalized amphiphilic polymeric nanoparticles with the advantage of "Green appeal" further enhance its applications as an important strategy.

These unique alternating copolymer micellar **nanoparticles** have been used successfully for the encapsulation of a large number of drugs of different classes and delivery vehicles targeted to human cancer cells.

A novel nanotechnology platform for in vivo imaging and delivery of multifunctional therapeutics of cancer has also been designed based on perfluorinated amphiphilic copolymers. These **nanoprobes** are highly unique because of their ability to image and treat the cancer tumors by delivering the drugs to the cancer tumor sites. Recently, we have synthesized cationic polymers and **dendrons** that constitute of quaternary ammonium functional groups, polyglycerol and poly (ethylene glycol) units. The positive charge introduced on these macromolecules forms the basis for specific interactions between ligand and receptor or enzyme and substrate, i.e., as ammonium cations, they may bind to polyanionic DNAs and also to negatively charged cell surfaces to trigger endocytosis. Thus, they may serve as gene siRNA delivery vehicles in order to cure many hereditary diseases and treat acquired diseases resulting from either multigenic disorders or foreign viral genes.

Further, highly useful novel, non-toxic "environment-friendly" non-halogenated flame-retardant organosilicon polymeric materials and nanocomposites have been developed using the above environmentally benign "green" biocatalytic technologies. These show superior properties than commercial flame-retardant materials.

These results shall be presented in the talk.

References

1. Kumar A, Khan A, Malhotra S, Mosurkal R, Dhawan A, Pandey MK, Singh BK, Kumar R, Prasad AK, Sharma SK, Samuelson LA, Cholli AL, Len C, Richards NGJ, Kumar J, Haag R, Watterson AC, Parmar VS (2016) Synthesis of macromolecular systems via lipase catalyzed biocatalytic reactions: applications and future perspectives. Chem Soc Rev. https://doi.org/10.1039/c6cs00147e
2. Colton C, Watterson A, Parmar VS, Fisher R, Kumar R, Kumar J (2015) Amphiphilic polymers and methods of use thereof. US Patent 8349991, 24 July 2015
3. Watterson AC, Mosurkal R, Kumar J, Parmar VS, Samuelson LA, Kumarand R, Tucci VB (2015) Mixed organic and inorganic polymers. US Patent 9156945, 13 October 2015
4. Khan A, Sharma SK, Kumar A, Watterson AC, Kumar J, Parmar VS (2014) Novozym 435—catalyzed synthesis of polyesters and polyamides of medicinal and industrial relevance. Chem Sus Chem **7**:379
5. Malhotra S, Bauer H, Tschiche A, Staedtler AM, Mohr A, Calderón M, Parmar VS, Hoeke L, Sharbati S, Einspanier R, Haag R (2012) Glycine-terminated dendritic amphiphiles for nonviral gene delivery. Biomacromolecules 13:3087
6. Pandey MK, Tyagi R, Yang K, Fisher RJ, Colton CK, Kumar J, Parmar VS, Aiazian E, Watterson AC (2011) Design and synthesis of perfluorinated amphiphilic copolymers: smart nanomicelles for theranostic applications. Polymer 52:4727
7. Malhotra S, Calderon M, Prasad AK, Parmar VS, Haag R (2010) Novel chemoenzymatic methodology for the regioselective glycine loading on polyhydroxy compounds. Organic Biomol Chem 8:2228
8. Gupta S, Pandey MK, Levon K, Haag R, Watterson AC, Parmar VS, Sharma SK (2010) Biocatalytic approach for the synthesis of glycerol-based macroamphiphiles and their self-assembly to micellar nanotransporters. Macromol Chem Phys 211:239
9. Mosurkal R, Samuelson LA, Parmar VS, Kumar J, Watterson AC (2007) Biocatalytic synthesis of organosiloxane copolyimide. Macromolecules 40:7742
10. Kumar R, Watterson AC, Parmar VS, Kumar J, Samuelson LA. Enzymatic synthesis of polymers. US Patent 6962963, 8 November 2005
11. Kumar R, Cheng M-H, Parmar VS, Samuelson LA, Kumar J, Nicolosi R, Yoganathan S, Watterson AC (2004) Supramolecular assemblies based on copolymers of PEG600 and functionalized aromatic diesters for drug delivery applications. J Am Chem Soc 126:10640
12. Kumar R, Tyagi R, Parmar VS, Samuelson LA, Kumar J, Schoemann A, Westmoreland PR, Watterson AC (2004) Biocatalytic synthesis of highly flame retardant inorganic-organic hybrid polymers. Adv Mat 16:1515; Science 306:375
13. Kumar R, Bruno F, Parmar VS, Kumar J, Watterson AC, Chittibabu KG, Samuelson LA (2004) "Green"-enzymatic synthesis of pegylated phenolic macromer and polymer. Chem Commun 862
14. Kumar R, Sharma AK, Parmar VS, Watterson AC, Chittibabu KG, Kumar J, Samuelson LA (2004) Flexible, dye-sensitized nanocrystalline solar cells employing biocatalytically synthesized polymeric electrolytes. Chem Mat 16:4841
15. Sharma SK, Kumar R, Kumar S, Mosurkal R, Parmar VS, Samuelson LA, Watterson AC, Kumar J (2004) Influence of EDA-π interactions in drug encapsulation using nanospheres. Chem Commun 2689

Prof. Ashok K. Prasad

Department of Chemistry
University of Delhi
Delhi, India

Sugar-Modified Nucleosides, Amphiphiles, and Pseudorotaxanes of Importance

Ashok K. Prasad*
Department of Chemistry, University of Delhi
New Delhi-110 007, India
Email: ashokenzyme@gmail.com

The use of biocatalysts in the modification of sugars has become an attractive alternative over conventional chemical methods due to their selectivity and high efficiency. We have successfully used lipases for the synthesis of sugar-modified bicyclic nucleosides. Further, we have used the modified sugar precursor for the synthesis of amphiphiles, chiral crown ether analogs, and derived [2] pseudorotaxanes.

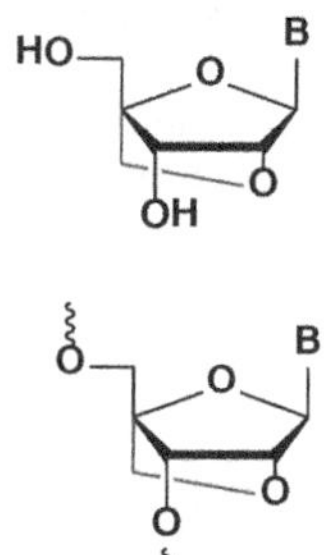

LNA monomers

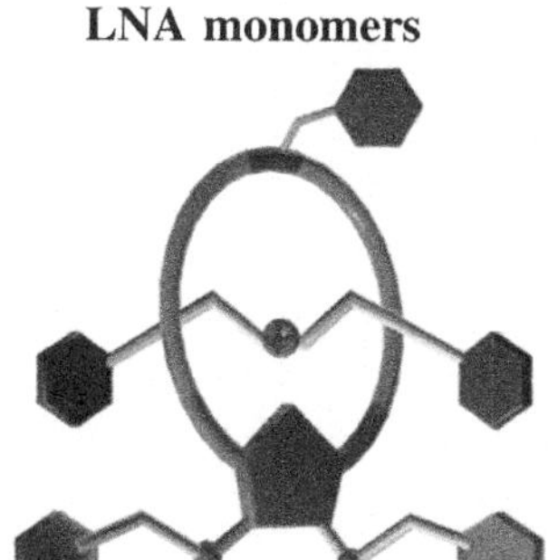

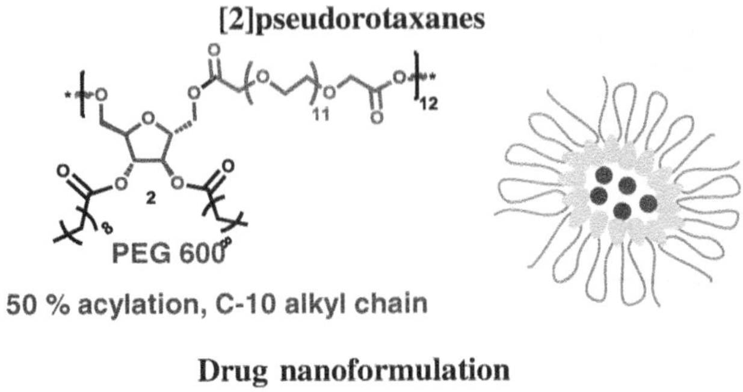

References

1. Prakash TP (2011) Chem Biodivers 8:1616
2. Paquette LA (2004) Aust J Chem 57:7
3. Singh SK, Sharma VK, Olsen CE, Wengel J, Parmar VS, Prasad AK (2010) J Org Chem 75:7932
4. Singh SK, Sharma VK, Bohra K, Olsen CE, Prasad AK (2011) J Org Chem 76:7556
5. Sharma VK, Singh SK, Bohra K, Chandrashekhar Reddy L, Khatri V, Olsen CE, Prasad AK (2013) Nucleosides, Nucleotides & Nucleic Acids 32:256
6. Sharma VK, Kumar M, Olsen CE, Prasad AK (2014) J Org Chem, 79:6336
7. Sharma VK, Rungta P, Prasad AK (2014) RSC Adv 4:6336
8. Kumar M, Sharma VK, Olsen CE, Prasad AK (2014) RSC Adv 4:37231
9. Sharma VK, Kumar M, Sharma D, Olsen CE, Prasad AK (2014) J Org Chem 79:8516
10. Rangarajan TM, Singh R, Brahma R, Devi K, Singh RP, Singh RP (2014) Chem Eur J 20:14218
11. Srivastava S, Singh SK, Singh VK, Mangala P, Prasad AK (2015) Nucleosides, Nucleotides, Nucleic Acids 34:399
12. Rangarajan TM, Brahma R, Ayushree, Prasad AK, Verma AK, Singh RP (2015) Tet Letts 56: 2234
13. Rangarajan TM, Devi K, Ayushree, Prasad AK, Singh RP (2015) Tetrahedron 71:8307
14. Khatri V, Kumar A, Singh B, Malhotra S, Prasad AK (2015) J Org Chem 80:11169
15. Singh A, Khatri V, Malhotra S, Prasad AK (2016) Carbo Res 421:25
16. Kumar M, Kumar R, Rana N, Prasad AK (2016) RSC Adv. https://doi.org/10.1039/C5RA25222A

Prof. Xi Chen

Department of Chemistry
University of California-Davis, USA

Using Enzymes for Efficient Production of Carbohydrates

Xi Chen*
Department of Chemistry, University of California-Davis
One Shields Avenue, Davis, CA 95616, USA

Cells from different organisms are all coated with carbohydrates which play important roles in biology. They are directly involved in cell–cell interaction, signaling, inflammation, immune regulation, cancer metastasis, as well as bacterial and viral infection. Due to their important functions, carbohydrates are attractive synthetic targets and have inspired efforts for developing carbohydrate-based therapeutics. Nevertheless, carbohydrates are challenging compounds for chemical synthesis. We take advantage of enzymes, nature's highly efficient and selective catalysts, and use them with the combination of chemical synthesis to obtain structurally complex carbohydrates. Several one-pot multienzyme (OPME) chemoenzymatic synthetic systems have been developed for producing a diverse array of carbohydrates which are being used as probes and reagents to elucidate the functions of carbohydrate-binding proteins, to identify carbohydrate-binding antibodies as cancer markers, and to develop carbohydrate-derived therapeutics. The use of enzymes simplifies synthetic procedures and minimizes the use of organic solvent. This effort is one of the many examples of multidisciplinary research activities carried out in the Department of Chemistry at the University of California-Davis.

Prof. K. C. Gupta

Polymer Research Laboratory
Department of Chemistry
Indian Institute of Technology Roorkee, India

Polymer-Supported Green Catalysts for Oxidation Reactions

K. C. Gupta*
Polymer Research Laboratory, Department of Chemistry
Indian Institute of Technology Roorkee, Roorkee-247667, India
Email: guptakc03@yahoo.co.in, Tel: 01332 285325

Transition metal ions complexes have shown remarkable catalytic activity in various oxidation reactions in comparison with other metal ions complexes. The transition metal ions complexes are found to be attractive for catalyzing reactions at room temperature in various industrial processes. The anchoring of metal ions complexes on various insoluble supports has shown a significant improvement in their catalytic activities and selectivity. The catalytic activity of metal ions complexes has been studied on various supports such as alumina, clays, and zeolites and found to be highly active in comparison with unsupported transition metal ions complexes. Though catalysts have shown high activities on these supports, still these inorganic supports have some limitations; hence, the anchoring of transition metal ions complexes on nano-sized cross-linked polymer supports has shown better results and applicability in under different conditions. The shapes and forms of polymer supports provide more opportunities in controlling the activity and selectivity, which is usually difficult to be achieved with conventional inorganic supports. The architectural control of polymeric supports, and porosity has contributed significantly in controlling the activity of Schiff base complexes of transition metal ions in oxidation reactions. The microenvironments provided by polymer support have played a significant role in providing effective interactions between catalysts and reactants to achieve products of desired selectivity and yield. The talk will cover various issues of green chemistry and environment-friendly oxidation reactions using polymer-supported transition metal complexes.

Dr. Beer Singh

Associate Director
Defence R&D Establishment
Gwalior, India

Ionic Liquids in CBRN Defence

Beer Singh*
Defence R&D Establishment, Gwalior, India

Ionic liquids (ILs) are salt that exist in the liquid phase at and around 298 K which represent a new class of solvents with non-molecular ionic character. ILs have attracted great attention as a replacement for traditional organic solvents as they possess many attractive properties such as intrinsic ion conductivity, low volatility, high chemical and thermal stability, low combustibility, and wide electrochemical windows. The ability to tune an appropriate IL based on specific application makes IL also as "designer solvents." The entire ionic composition, intrinsic conductivity, large electrochemical potential window, and negligible vapor pressure made ILs as greener sensing media to use in the development of stable electrochemical sensing methodologies for gaseous analytes when addressing the health and safety concerns associated with many organic solvent applications. Due to negligible volatility of these solvents, they are considered "greener" for the environment as they do not evaporate like volatile organic compounds (VOCs). ILs have been widely used in electrodeposition, electrosynthesis, electrocatalysis, electrochemical capacitor, lubricants, solvents to manufacture nanomaterials, extraction, gas absorption agents, and so forth. The non-volatility and intrinsic catalytic property of IL have also been explored for the decontamination of chemical warfare agent (CWA) and in the same line wide electrochemical potential window and catalytic nature of IL have been used for electrocatalysis of CWA. Certain room temperature ionic liquids (RTILs) can work under any field condition as sensing medium for years together due to the low volatility and negligible vapor pressure when compared to molecular solvents offers the opportunity to use this methodology in field condition without environmental humidity effect. The IL-based detection and decontamination strategy is the first step in developing a new, easy-to-apply product that could be used by the first responders to detect and decontaminate the CWA in the field and could provide a new paradigm shift in detection and decontamination approach in a greener and environmental sustainable way.

Prof. A. K. Narula

Director
Centre of Excellence in Pharmaceutical Sciences (CEPS)
Guru Gobind Singh Indraprastha University
Delhi, India

Effect of Different Polymerization Techniques on the Surface Morphology of Polypyrrole—A Scanning Electron Microscopy Study

Shruti Peshoria, Anudeep Kumar Narula*
Molecular Chemistry Laboratory
University School of Basic and Applied Sciences
Guru Gobind Singh Indraprastha University
Sector 16 C, Dwarka, New Delhi-110078, India
Corresponding Author. Tel: +91-11-25302423
Fax: +91-11-2530111
Email: aknarula58@gmail.com, researchchemlab58@gmail.cpm

Energy harvesting directly from sunlight using solar cell technology is the most promising way to utilize the renewable source to address the global energy demand. Nowadays, polymeric solar cells are receiving enormous attention. The most critical challenge is the designing of solar cells so that they should possess strong absorption ability, high hole mobility, and suitable HOMO–LUMO levels. Molecular engineering and synthesis of these conjugated polymers play a vital role in determining their photovoltaic characteristics. Conducting polymer polypyrrole (PPy) was prepared by three different polymerization methods viz. electrochemical polymerization (EP), interfacial polymerization (IP), and chemical oxidative polymerization (COP) to study the change in morphology with scanning electron microscopy (SEM). The samples were analyzed by Fourier transform infrared spectroscopy (FTIR) to confirm the successful formation of ppy with the appearance of characteristic bands at 3424, 1575, 1049, and 964 cm^{-1}. Optical studies were done by UV–Vis absorption spectroscopy that displayed $\pi \rightarrow \pi^*$ and polaronic/bipolaronic transitions of ppy. Examination of SEM and HRTEM micrographs gave an insight into different morphologies of PPy.

Keywords Polypyrrole · SEM · Cyclic voltammetry · Interfacial polymerization Morphology · Characterization

Dr. Charu Kapil Midha

Scientific Coordinator
Science and Technology Department
French Embassy to India, New Delhi, India

Indo-French S&T cooperation and Mobility of Ph.D. Students

Charu Kapil Midha*
Science and Technology Department
French Embassy to India, New Delhi, India
Email: charu.kapil@diplomatie.gouv.fr; ckp@ifindia.in

France and India share a long tradition of cooperation in Science and Technology. Sharing the same vision and joining hands to address the new challenges of our societies over the time, France is always committed to strengthen the existing fruitful and dynamic partnership. It offers tremendous opportunities for both our scientific communities to develop added-value cooperation in any domain from mathematics to social sciences and humanities. France is the fifth partner of India in terms of joint scientific publications. Based on a rich and varied research landscape, France benefits from numerous reputed higher education and research institutions which are actively involved in Indo-French scientific collaborations.

To promote the Indo-French cooperation in S&T, there are several programs and tools available; among them, the first one is the CEFIPRA (Indo-French Centre for the Promotion of Advance Research). This unique bilateral body, co-funded by the two governments, offers grants for collaborative research projects both academic and industrial, which include fellowship for master's, Ph.D., and postdoc students. Other dedicated program managed by the CEFIPRA provides two-way mobility support to Ph.D. students (Raman-Charpak Fellowship), allowing Ph.D. scholars from France and India to spend 2–6 months in other countries during their thesis. In addition, MoUs, joint research laboratories, the Research, Development and Innovation Club (RD&I), and the Indo-French Water Network (IFWN) to name a few form the strong Indo-French cooperation network in S&T established over the last decades.

In establishing such a strong and sustainable network between India and France at industrial and academic level, CEFIPRA plays a major role. Apart from collaborative research projects, several Indo-French seminar and workshops are also sponsored by CEFIPRA leading to strong networking and resulting into new collaborations and added value to the existing network.

For higher studies, France is the choice among Indian for its excellent offers and opportunities. Around 73 universities that exist in France comprise of 273 doctoral schools with nearly 2500+ dedicated and specialized laboratories for higher studies, in particular École doctorale de Physique—Adum, Doctorat de l'Université Grenoble Alpes-École Doctoral de Physique, École Doctoral de Physique de la Region Parisienne, Université de Strasbourg: École Doctorale Physique et Chimie, etc., from different reputed universities dedicated to chemistry and physics.

Dr. Rakeshwar Bandichhor

Director API—R&D
Dr. Reddy's Laboratories
Hyderabad, India

DKR-Based Process Development of Clopidogrel Bisulfate

Dattatray Metil, Vilas H. Dahanular and Rakeshwar Bandichhor*
Innovation plaza, Integrated Product Development
R&D, Dr. Reddy's Laboratories Ltd., Survey Nos. 42, 45, 46 and 54
Bachupally, Qutubullapur, R. R. District-500090, A.P., India
Email: rakeshwarb@drreddys.com

Kinetic resolution is one of the most effective strategies to obtain enantiomerically pure compounds. However, this strategy is limited to a maximum throughput not more than 50%. Rest of the 50% is considered waste unless it is recycled. Recycling would also involve processing waste and cost which would lead overall process inefficient and nongreen. In order to overcome these limitations, dynamic kinetic resolution (DKR) which is a process of racemization combined with resolution is innovatively devised. In order to obtain equilibrium between the enantiomers, conversion of the slow-reacting enantiomer into the fast-reacting one takes place.

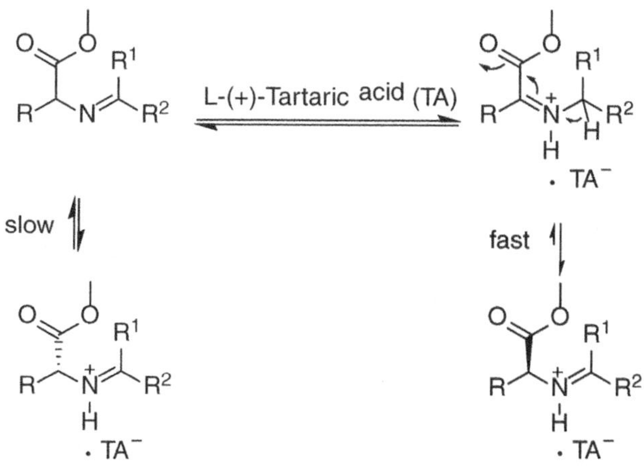

Fig. 1 DKR involving lipophilic ketone and L-(+)-tartaric acid

The fast converting enantiomer in the form of diastereomeric salt is therefore never exhausted, and the process will afford a theoretical yield of 100% of the desired enantiopure product after workup. In our endeavor, we employed DKR strategy in the resolution of one of the intermediates of clopidogrel which is a thienopyridine class of antiplatelet agent meant for inhibiting blood clots in coronary artery, peripheral vascular and cerebrovascular diseases, thereby helping in preventing heart attack and stroke.

In order to develop a high yielding robust process, we opted for more lipophilic ketone and L-(+)-tartaric acid and this combination found to play critical role in DKR (Fig. 1). Moreover, statistical design of experiments (DoE) is used to optimize a process for the preparation of clopidogrel bisulfate. In this work, the application of quality by design (QbD) principles to the development of three important stages of API is described.

Mr. Abhishek Gupta
Director
Clean India Ventures—Graphisads Pvt Ltd.
New Delhi, India

Green Waste Reprocessor

Abhishek Gupta*
Clean India Ventures—Graphisads Pvt Ltd.
4/24A, Asaf Ali Road, Near Delhi Gate, New Delhi-110002, India

Sixty-five million tonnes of waste is generated by 350 million urban population every year. Ninety percentage of this waste goes to landfills that have huge implications such as contaminated air, water in its surroundings, higher likelihood of epidemic, unhygienic surroundings, and a significant direct and indirect cost. Sixty percentage of the waste generated is organic in nature and can be recycled before it becomes waste.

While a lot of big centralized projects have come up in India, there is a requirement of decentralized technology to get rid of the waste and get a beneficial use of that waste by converting it into a valuable by-product.

Green Waste Reprocessor (GWR) converts all the organic waste (green, food/vegetable, temple waste) into a valuable by-product without creating pollution and affecting the environment. While there are various players who are in the field of waste to compost, we have the added advantage of creating multiple by-products from different types of waste. For example, we create Organic Havan Samagri from flower waste, therefore giving it back to the temples the Havan Samagri for their further use.

It is a decentralized machine working with the help of electricity, and they can be installed at the source of waste only to decrease the cost of transportation of waste to landfill sites.

We have different by-products for different types of waste. For green waste, we create organic compost and fuel sticks. Organic compost can be used in agriculture, farming, plantations, and fuel sticks can be used in clean cooking. They are smokeless pellets which are used for burning, and they have a special cooking stove in which they are used.

In a decentralized technology, which is the need of the hour, any city can use approximately 100 machines and create a system to reprocess more than 60% of its organic waste. Even if a few machines are under maintenance, waste can be shifted to other machines. With a centralized project, any big maintenance issue can have an adverse impact on city.

We save on fuel as our technology helps reduce transportation of waste, as well as we save land which is one of the most precious resources by reducing landfill requirement.

This project will cover most of the unskilled labor, and people below the poverty line since this machine can generate employment for all the ragpickers, beggars. They would be employed to pick the waste and keep it near the machine to process the waste. Before putting the waste in the machine, employees will be asked to segregate the waste and only

We shall also look forward to create Enviropreneurs who will be responsible for operations of the machine and selling the by-product in the market. We shall also look to engage self-help group of women which are keen to work for the environment. This is in line with the Government's initiatives of Empowering Women.

Our project is in sync with key initiatives of Central Government, i.e., Swachh Bharat, Smart City, Make In India, Skill India, Startup India, etc.

Contents

About the Editors

Prof. Virinder S. Parmar obtained his B.Sc. Honors, M.Sc., and Ph.D. from the University of Delhi and did his postdoctoral research at Cornell University, Harvard University, the University of Basel, and Imperial College. He has been Faculty Member at the University of Delhi for 44 years and has served as Head of the Department of Chemistry and Chairman of Board of Research Studies. He has been Visiting Professor at the University of Massachusetts Lowell, University of Southern Denmark, Institute of Advanced Sciences, and the Central University of Haryana. His research interests include nanoscience, synthetic organic chemistry, biocatalysis, polymer chemistry, nucleic acid chemistry, green chemistry, advanced materials, and chemistry of natural products. He has mentored 85 Ph.D. and postdoctoral scientists and has published 490 research papers in addition to being co-inventor on 11 patents. He is Member of the IUPAC Organic and Biomolecular Chemistry Division's Subcommittees on Biomolecular Chemistry and Green Chemistry.

Dr. Priti Malhotra obtained her Ph.D. from the University of Delhi and is a Senior Chemistry Faculty at Daulat Ram College, University of Delhi, for the last 25 years. She has a vast experience of teaching inorganic chemistry to undergraduate students and has also organized various workshops, seminars, and lectures. Her research interests include green chemistry, nanobiotechnology, and material science. She has mentored 40 undergraduate and 5 Ph.D. students and has published 30 research papers in prestigious scientific journals and various national and international conference proceedings. She has also authored a book on "Analytical Chemistry."

Dr. Divya Mathur received her B.Sc. (2006), M.Sc. (2008), and Ph.D. (2013) in Chemistry from the University of Delhi, India, and did her postdoctoral research at Ghent University, Belgium. She has worked as Visiting Research Scholar at the Acadia University, Canada. She is a recipient of Erasmus Mundus Postdoctoral

Scholarship by European Union, Canadian Commonwealth Scholarship by Government of Canada, and National Postdoctoral Fellowship by Government of India. Presently, she is Assistant Professor at Daulat Ram College, University of Delhi, India. She has published over 25 peer-reviewed papers in reputed international journals. Her research interests include nanoscience, synthetic organic chemistry, biocatalysis, nucleic acid chemistry, and green chemistry.

Bioremediation of Xenobiotics: An Eco-friendly Cleanup Approach

Alok Bharadwaj

Abstract Xenobiotics are the chemicals, which are not found in nature and are said to be foreign substances in the atmosphere. These compounds are synthesized by human beings e.g. pesticides, that may be added in the soil to kill the harmful pests but in addition this, these pesticides also kill the beneficial microorganisms that are responsible for the fertility of soil. Moreover, these compounds can be accumulated in food chain and cause harm to the flora and fauna of such ecosystem. For the degradation of such xenobiotic compounds various physico-chemical and biological methods have been used but all these methods produce toxic by-products that are hazardous to the environment. Thus Bioremediation is a promising tool for the degradation of such compounds. In the present paper, we have emphasized on the remediation of xenobiotics by using microorganisms. Bacteria and Fungi play an important role in breaking down certain hazardous substances into simpler fragmented forms. Bacteria, which are used for bioremediation of xenobiotics includes aerobic, anaerobic, Methanotrophic, Methanogenic bacteria, Cyanobacteria and Sphingomonads. *Pseudomonas* species has been used for the degradation of several xenobiotic compounds. In addition to this, certain fungi such as *Aspergillus, Rhizopus, Botrytis, Neurospora* etc. have been used for the heavy metal biosorption. Hence we can say that fungi are one of the promising tools for the eco-friendly degradation of xenobiotic.

Keywords Xenobiotics · Microorganisms · Bioremediation

1 Introduction

The term 'bioremediation' means transformation of a chemical compound from highly complicated form (organic) to simple (inorganic) form through biological means. If we say a compound is biodegradable, it means that it can be converted

A. Bharadwaj (✉)
Department of Biotechnology, GLA University, Mathura, UP, India
e-mail: alok.bhardwaj@gla.ac.in

V. S. Parmar et al. (eds.), *Green Chemistry in Environmental Sustainability and Chemical Education*, https://doi.org/10.1007/978-981-10-8390-7_1

into various inorganic forms or can be mineralized i.e., possible to convert into carbon dioxide and water.

Xenobiotics are the chemical compounds, which are manufactured by human and their degradation is not possible. These are foreign compounds, when collected in the atmosphere pose dangerous effects on the atmosphere [1–3]. Examples of xenobiotics are pesticides, fuels, solvents, alkanes, polycyclic hydrocarbons (PAH's), antibiotics, synthetic azo dyes, pollutants, polyaromatic, chlorinated and nitro aromatic compounds [4]. Moreover, certain compounds are recalcitrant i.e. the compound are resistant to being broken down through chemical process because of the presence of halogen, nitro or sulphonyl groups.

The major examples of xenobiotics are DDT and halogenated aromatic compounds because they pose hazardous impact on the atmosphere. Although, many such compounds are highly beneficial to the humans but their long term presence in the environment cause deleterious effects.

Recalcitrant toxic pesticides such as BHC, PCBs, and DDT are not easily biodegraded and their concentration goes on increasing in the soil and water with time. For example, there is a continuous increase in the concentration of DDT at successive trophic levels in food chain. Model compounds of DDT such as diphenymethane and dichlorodiphenylmethane are biodegrade by various microbes. *Pseudomonas putida* converts DDT to several compounds under anaerobic conditions. In aerobic conditions, ring cleavage yield p-chlorophenylacetic acid that in turn served as a substrate for an *Arthrobactor* species. A number of reviews have been available on the bioremediation of xenobiotics [4–7].

2 Basis of Bioremediation

Microbes play the most important role in the process of biodegradation. Certain abiotic mechanisms and photo-oxidation also play an important role in the degradation of certain organic chemicals but such transformations are generally incomplete because these processes cannot convert the compound into inorganic form (Fig. 1).

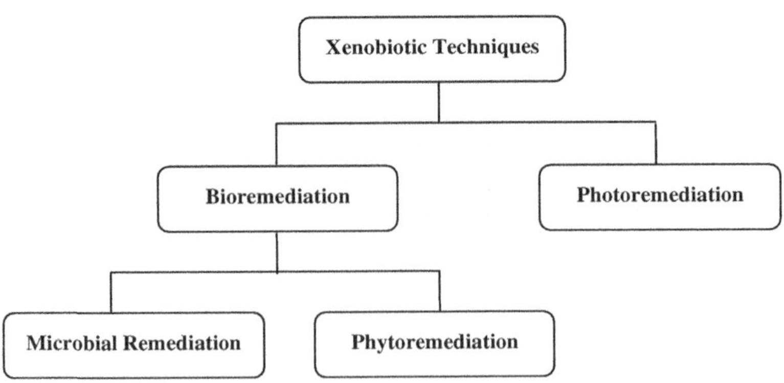

Fig. 1 Classification of different xenobiotic compounds

Before the industrialization, biosphere of earth remained constant because of more or less balanced biosynthesis and biodegradation reactions. During the course of evolution, a vast variety of chemical compounds are biosynthesized in nature and microbes were exposed to these compounds. For this millions of year's exposure, they have developed the capacity and mechanism to attack these compounds. Out of several chemical compounds synthesized by chemists, many have structural features and bonding similar to that of natural compounds, so can be biodegraded.

The xenobiotic compounds having recalcitrant characteristics have following features

(i) These compounds are not recognized by the microorganisms as substrate. So, can't be degraded by them.
(ii) For their transportation into the microbial cell, they do not have permease enzyme.
(iii) These are complex molecules. So, very difficult to enter inside the microbial cell.
(iv) These compounds are highly stable in the nature. In addition to this they are insoluble in water.
(v) Xenobiotic compounds are very poisonous in nature.

On the basis of chemical composition, the recalcitrant xenobiotic compounds can be divided into the following types (Fig. 2).

2.1 *Halocarbons*

These compounds are mainly used in the preparation of pesticides, insecticides etc. These are volatile, when free into the environment cause damage to the ozone layer and when gathered in soil, leading to biomagnifications.

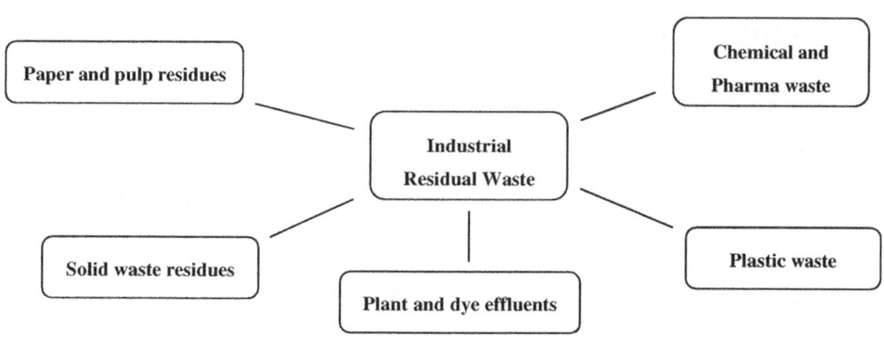

Fig. 2 Types of recalcitrant xenobiotic compounds

2.2 Polychlorinated Biphenyls (PCBs)

These compounds are inert in nature and are used in the manufacturing of plasticisers, insulator coolants in transformers etc.

2.3 Synthetic Polymers

These are high molecular weight compounds, which are insoluble in water and used in the manufacturing of plastics like polymer, polyvinylchloride.

2.4 Alkylbenzyl Sulphonates

These compounds are not degraded by microorganisms because of the presence of sulphonate group. They are used in the manufacturing of detergents.

2.5 Oil Mixtures

Due to accidental oil spilling in the ocean, oil spread over the water surface and its breakdown by microorganisms results ineffective. Oil is insoluble in water. So, it becomes recalcitrant.

From this discussion, it has been clear that the recalcitrant characteristic of a xenobiotic compound is directly connected with its complexity and it is also noted that recalcitrant property is enhanced as their complexity increased.

3 Hazards from Xenobiotic Compounds

The impacts of xenobiotic compounds are very hazardous in nature. Their dangerous effect is not only on lower eukaryotes but higher eukaryotes are also affected. They are very toxic to human beings also. Their exposure may cause certain skin diseases and prolong exposure may cause cancer also. Because they posses recalcitrant property, they are not degraded in the nature and may remain persist for long time in the environment may result in the form of bioaccumulation or biomagnifications. In bioaccumulation or biomagnifications, they enter into the food chain and their concentration will increase with the increase in the trophic level of that ecosystem.

4 Removal of Xenobiotics

Unluckily, xenobiotics are used widely among the modern society. To survive the better life, many industries—manufactured different chemical compounds but majority of them are poisonous and toxic for the living beings. e.g. the use of plastics, pesticides, paints, textiles and pharmaceuticals can not be eradicated from our daily needs but now the time has come that we should aware to overcome this huge problem and protect our future from the curse of xenobiotics. One of the most efficient tools to get rid of this problem is the use of microorganisms that may degrade these toxic chemicals from the atmosphere. Because microorganisms are easily grow and multiply in laboratory. So, they can be a useful device for the degradation of such chemical compounds. The enzymes present in these microorganisms are able to degrade these chemical compounds. The enzymes present in these microbes are able to degrade these chemicals into harmless end products. Different types of microorganisms are used to degrade different type of waste but unfortunately there are some artificially human made chemical compounds, which are not degraded by the microorganisms because they are not able to break certain chemical bonds present in the chemical compounds.

One of the recent branches of the science i.e. genetic engineering makes it possible to construct such type of microorganisms, that able to degrade these xenobiotic compounds. Among such microorganisms *Pseudomonas* genus are mostly employed for their degradation. Such bacteria have extra chromosomal DNA i.e. TOL plasmid, that have the capacity to degrade toluene and xylene.

There is one more approach to degrade these xenobiotic compounds by using the combined efforts of human and microbial activity. In this technique the recalcitrant chemicals are first break into smaller parts and then subjected to microbial degradation. An example is the accidental oil spilling, which is very dangerous for the life of aquatic organisms. By the combining efforts of human and microbes, aquatic organisms may survive but if the area of oil spill is very vast, then the rate of oil degradation by microbes becomes very slow due to the toxicity of oil. But now, recently a chemical is used for solving this problem.

A chemical name SOT II (solid oil treatment) is an inorganic solid absorbent, which is chemically inert. Whenever a oil spill occurs, this chemical is spread over the oil spill area. Since it is a strong absorbent, it will compact in small particles with oil and sink at the bottom of the sea. Due to the property of strong absorbent SOT II particles can't leave the oil spill. As the superficial layer of oil has removed from the water, the living organism survives as they come in direct contact with oxygen and sunlight. The particles of SOT II with oil, which sinks at the bottom of sea will become the food of for oil degrading organisms i.e. bacteria, algae, fungi and protozoa. This process takes place rapidly at the bottom of the sea and within a week these oil degrading organisms degrade the oil with the production of end products i.e. water and CO_2.

This technique shows 99% success in the removal of oil from the sea and its degradation and elimination from the ecosystem.

As we all know, xenobiotics are the toxins and detoxification is a successful remedy to neutralize these toxins. In detoxification process, the negative impact of these toxins can be minimized by planning the detox diet. The main objective of detox diet is to eradicate these toxins because this diet contains 40% solid and 60% liquids. Solid food contains certain phytochemicals e.g. indole, flavanoids etc. e.g. pineapple contains bromelain. Papaya contains papain, which is a colon cleaner and improves immune system. Consumption of raw food such as sprouts, dried fruits, nuts, milk etc. are also the examples of detoxification because these foods contains antioxidants that enhance the immune system of our body.

5 Microbial Degradation of Xenobiotics

To overcome the needs of increasing population, the use of pesticide in modern agriculture is increasing day by day but the results are very dangerous because these pesticides persist in the ecosystem and don't undergo biological transformation (Table 1).

In the degradation of xenobiotics compounds, the role of microorganisms is very important because these microorganisms degrade the xenobiotics into the simpler or non hazardous compounds. Among these microbes, bacteria have the noble detoxifying abilities. Examples of aerobic and anaerobic xenobiotics degradative

Table 1 Duration of persistence of insecticides and herbicides in soil

Biocides	Time taken for 75–100% disappearance
A. *Chlorinated insecticides*	
DDT (l,l,l-trichloro-2,2-bis-(p-chlorophenyl) ethane)	4 years
Aldrin	3 years
Chlordane	5 years
Heptachlor	2 years
Lindane (hexachloro-cyclohexane)	3 years
B. *Organophosphate insecticides*	
Diazinon	12 years
Malathion	1 week
Parathion	1 week
C. *Herbicides*	
2,4-D (2,4-dichlorophenoxyacetic acid)	4 weeks
2,4,5-T	30 weeks
Atrazine	40 weeks
Simazine	48 weeks
Propazine	1.5 years

Source [8]

bacteria are *Pseudomonas, Gordonia, Bacillus, Moraxella, Micrococcus, Escherichia, Sphingobium, Pandoraea, Rhodococcus*, and anaerobic xenobiotics degradative bacteria are *Pelatomaculum, Desulphovibrio, Methanospirillum, Methanosaeta desulfotomaculum, Syntrophobacter, Syntrophus* [6, 9]. Among them, *Pseudomonas* species and *Bacillus* species have been the most widely studied. Many other bacterial species which assist in degradation of recalcitrant xenobiotic compounds are listed in Table 2.

Table 2 List of xenobiotic compounds degraded by microorganisms

Xenobiotic compounds	Microorganisms degrading xenobiotics	Isolated sites	References
Petroleum hydrocarbons: crude oil, engine oil, petrol, diesel, brake oil	*Bacillus* sp. S6 and S35	Soil samples from storage and distribution centre of oil products in Tehran refinery and Siri Island	[10]
	Pseudomonas sp., PSI, PSII, PSIII	Soil samples from pil production site of ONGC (Lingal oil field project) and from local areas in Haridwar, India	[11]
	Pseudomonas putida (strain G1) and *Pseudomonas aeruginosa* (strain K1)	Soil samples from abandoned coal power plant (PHC) at Ijora-Olapa, Lagos	[12]
	Bacillus subtilis	Soil samples from automobile workshops and petrol bunks, Madurai, India	[13]
	Consortium 1: *Pseudomonas aeruginosa* strains S4.1 and S5 and *Bacillus* sp. strain S3.2 Consortium 2: Consortium 1 and *Bacillus* sp. strains 113i and O63 and *Micrococcus* sp. strain S	Bacterial strains purchased from centre for research in enzymes and microbiology, Malaysia	[14]
	Pseudomonas sp., *Vibrio* sp., *Bacillus* sp., *Corynebacterium* sp. and *Klebsiella* sp.	Soil samples from Missa Kaswal and Rajian oil fields, Gojar Khan and Chakwal districts	[15]
	Pseudomonas sp., *Vibrio* sp., *Bacillus* sp., *Corynebacterium* sp., *Flavobacterium* sp., *Micrococcus* sp. and *Morexella*	Soil samples from various gasoline and diesel spilled gas stations, Coimbatore, India	[16]
	Bacterial diversity	Soil samples from storage and distribution centre for oily products in south of Iran	[17]

(continued)

Table 2 (continued)

Xenobiotic compounds	Microorganisms degrading xenobiotics	Isolated sites	References
	Pseudomonas sp., *Acinetobacter* sp., *Bacillus* sp., *Corynebacterium* sp. and *Flavobacterium* sp.	Soil samples from automechanic workshops at Mgbuka-nkpor, Nigeria	[18]
	Pseudomonas aeruginosa PDKT-2, Bacillus licheniformis PDKT-5 and *marcescens PDKT-1 (HM998315*	Soil samples from auto mobile workshops, Pudukottai, Tamil Nadu, India	[19]
	Pseudomonas sp., *Arthrobacter* sp. and *Mycobacterium* sp.	Soil samples from garage area of heavy vehicles in Tehran-Saveh road, Iran	[20]
	Pseudomonas aeruginosa, Alcaligenes faecalis, Bacillus sp. and *Serratia* sp.	Soil samples from automechanic workshops at Ota, Western Nigeria	[21]
	Acinetobacter, Alcaligenes, Bacillus, Corynebacterium, Flavobacterium, Micrococcus and *Pseudomonas*	Higher purity n-alkane, cycloalkane, and aromatic HCs were obtained from Farmex Nigeria Limited, Sango-Otta	[22]
	Bacterial strains (RP1, RP2, RP4, RP12, DE5, DE7ii, DE8ii, OW13 and OW14)	Oil contaminated soil samples were collected from the oil production sites of Essar Oil LTD., Vadinar-Jamnagar and from Oil and Natural Gas Corporation (ONGC) Point (Gujarat-India)	[23]
	Micrococcus sp.	Soil samples were collected from automobile workshops in Mayiladuthurai	[24]
	Pseudomonas aeruginosa	Isolated from some crude oil flow stations' saver pit effluent in the Niger delta area of Nigeria	[25]
	Micrococcus and *Pseudomonas*	Soil samples contaminated with spent engine oil were collected from a mechanic workshop along Opopo gbooro, Iworoko Road, Ado-Ekiti	[26]
	Bacterial strains MJH1101, MJH1102, MJH1103, MJSH1104,	Soil samples were collected from two different oil contaminated sites: (A) Popular Service Garage, city station,	[27]

(continued)

Table 2 (continued)

Xenobiotic compounds	Microorganisms degrading xenobiotics	Isolated sites	References
		Lucknow, (B) Rajesh Garage, Jama masjid, Lucknow	
	Proteus vulgaris SR1	Bacterial strain was isolated from newly killed fish samples collected close to the point of spill in the Niger Delta region in Nigeria	[28]
	Pseudomonas sp., *Achromobacter* sp., *Bacillus* sp. and *Flavobacterium* sp.	Soil samples were obtained from a Judy Creek area of north-central Alberta, from a diesel fuel spill near Salmon Arm, British Columbia	[29]
	Flavobacterium sp., *Acinetobacterium calcoaceticum* and *Pseudomonas aeruginosa*	Soil samples collected from ENGEN, Amanzimtoti, South Africa	[30]
	Bacterial isolates SP4, SP5, SP6, SP7	Soil samples were collected from contaminated sites of Barmer near Mangala oil well, Rajasthan	[31]
	Arthrobacter, Halomonas, Pseudomonas, Bacillus, Klebsiella, Proteus	Soil samples from oil fields of ONGC sites located in Gujarat	[32]
	Pseudomonas alcaligenes LR14, Bacillus coagulans CR31, Klebsiella pneumonia CR23, Klebsiella aerogenes CR21 and Pseudomonas putrefacience CR33	Rhizosphere soil contaminated with spent engine oil in Sokoto, Nigeria	[33]
	Acinetobacter iwoffii, Aeromonas hydrophila, Pseudomonas aeruginosa, Pseudomonas putida	Soil samples from two oil refineries in Malaysia	[34]
Pesticides: Glyphosate	*Pseudomonas putida, P. aeruginosa and Acinetobacter faecalis*	Agricultural soil polluted with glyphosate in Osogbo Osun State, Nigeria	[35]

(continued)

Table 2 (continued)

Xenobiotic compounds	Microorganisms degrading xenobiotics	Isolated sites	References
Organochlorine-DDT	*Pseudomonas aeruginosa, P. putida, Stenotrophomonas maltophilia, Flavimonas oryzihabitans,* and *Morganella morganii*	Green bean coffee (Coffea arabica) from Veracruz was supplied by the Mexican Coffee Council	[36]
Tetrachlorvinphos	*Stenotrophomonas malthophilia, Proteus vulgaris, Vibrio metschinkouii, Serratia ficaria, Serratia* spp. and *Yersinia enterocolitica*	Tetrachlorvinphos contaminated agricultural soil, Mexico	[37]
Chlorpyrifos	Bacterial strains	Rhizospheric andnon-rhizospheric soil contaminated with organophosphorous pesticide	[38]
Organophosphorous pesticide - malathion	*Staphylococcus aureus*	Strain obtained from National Collection of Industrial Microorganisms, Pune	[39]
Atrazine	*Enterobacter* spp., *Bacillus* spp., *Providencia* spp. and *Pseudomaonas* spp.	Soil contaminated with atrazine in Egypt and Saudi Arabia	[40]
Organochlorine	Actinomycetes	Organochlorine contaminated soil	[41]

In the present scenario, the problem of xenobiotics remediation can be sorted out by the use of microorganisms that play a key role in the degradation of these xenobiotic compounds. An example of xenobiotics degradation is pentachlorophenol (PCP) that is a broad spectrum biocide, which has been used as fungicide, insecticide, herbicide, algicide, disinfectant and antifouling agent.

Bioreactors containing alginate immobilized along with Polyurethane foam immobilized PCP degrading *Flavobacterium* (ATCC39723) cells have been used to remove PCP from contaminated water. Absorption of PCP by Polyurethane immobilized matrix plays a role in reducing the toxicity of PCP (Fig. 3).

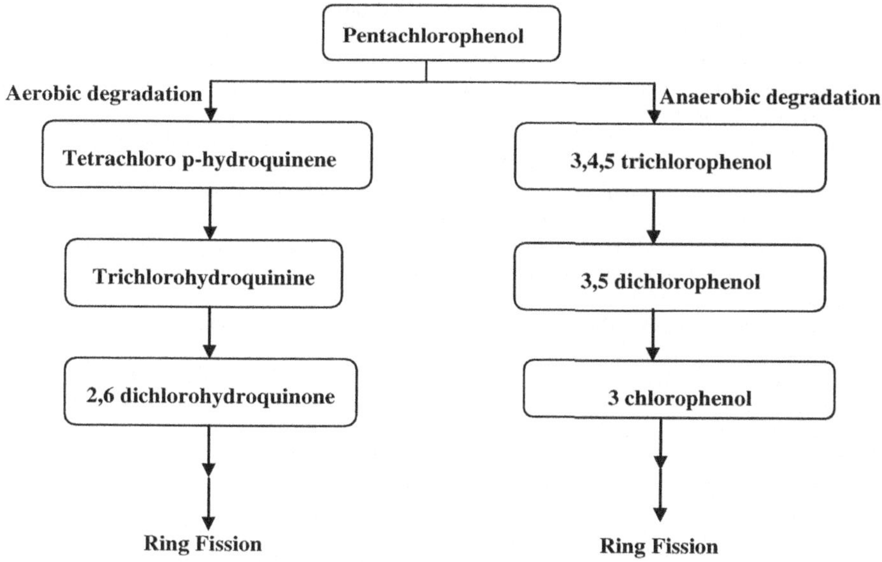

Fig. 3 Outline of aerobic and anaerobic degradation of pentachlorophenol

6 Conclusion

Xenobiotic compounds present a big problem in the environment but microorganisms play an important role in their bioremediation. From this paper, it has been clear that oil spilling is the world's greatest problem but with the help of genetically modified microorganisms, these oil spills can be cleaned up within few days. So, from this review, it may be concluded that microbial degradation is an important tool for the remediation of different xenobiotic compounds.

Acknowledgements Author would like to thanks Prof. A.K. Bhatia, HOD and Prof. Anoop Gupta, Director, Dept. of Biotechnology, GLA University to encourage during this work.

Conflicts of Interest Author declares no conflict of interest.

References

1. Gursahani YH, Gupta SG (2011) Decolourization of textile effluent by a thermophilic bacteria *Anoxybacillus rupiensis*. J Pet Environ Biotechnol 2:111
2. Sridevi V, Lakshmi MVVC, Swamy AVN, Rao MN (2011) Implementation of response surface methodology for phenol degradation using *Pseudomonas putida* (NCIM 2102). J Bioremed Biodegrad 2:121

3. Kumar KK, Prasad MK, Sarma GVS, Murthy CVR (2009) Removal of Cd(II) from aqueous solution using immobilized *Rhizomucor tauricus*. J Microbial Biochem Technol 1:015–021
4. Sinha S, Chattopadhyay P, Pan I, Chatterjee S, Chanda P, Bandyopadhyay D, Das K, Sen SK (2009) Microbial transformation of xenobiotics for environmental bioremediation. Afr J Biotech 8(22):6016–6027
5. Agarwal N, Shahi SK (2015) An environmental cleanup strategy—microbial transformation of Xenobioti compounds. Int J Curr Microbiol Appl Sci 4(4):429–461
6. Varsha YM, Naga Deepthi CH, Chenna S (2011) An emphasis on xenobiotic degradation in environmental cleanup. J Biorem Biodeg 11:1–10
7. Janssen DB, Dinkla IJT, Poelarends GJ, Terpstra P (2005) Bacterial degradation of xenobiotic compounds: evolution and distribution of novel enzyme activities. Environ Microbiol 7:1868–1882
8. Madigan MT, Martinko JM, Parker J (1997) Biology of microorganisms, 8th edn. Prentice Hall, Englewood Cliffs
9. Chowdhury S, Mishra M, Adarsh VK, Mukherjee A, Thakur AR, Chaudhuri SR (2008) Novel metal accumulator and protease secretor microbes from East Calcutta Wetland. Am J Biochem Biotechnol 4:255–264
10. Sepahi AA, Golpasha DI, Emami M, Nakhoda MA (2008) Isolation and characterization of crude oil degrading *Bacillus* spp. Iran J Environ Health Sci Eng 5(3):149–154
11. Mittal A, Singh P (2009) Isolation of hydrocarbon degrading bacteria from soils contaminated with crude oil spills. Ind J Exp Biol 47:760–765
12. Salam LF, Obayori OS, Akashoro OS, Okogie GO (2011) Biodegradation of bonny light crude oil by bacteria isolated from contaminated soil. Int J Agric Biol 13:245–250
13. Darsa KV, Thatheyus AJ, Ramya D (2014) Biodegradation of petroleum compound using the bacterium *Bacillus subtilis*. Sci Inter 20–25
14. Ghazali FM, Abdul Rahman RNZ, Salleh AB, Basri M (2004) Biodegradation of hydrocarbons in soil by microbial consortium. Int Biodeterio Biodegrad 54:61–67
15. Raza C, Billal A, Jahan N (2010) Evaluation of biodegradation potential of bacteria in crude oil contaminated soil. Biologia 5(1&2):77–85
16. Sathishkumar M, Binupriya AR, Balk S, Yun S (2008) Biodegradation of crude oil by individual bacterial strains and a mixed bacterial consortium isolated from hydrocarbon contaminated areas. Clean 36(1):92–96
17. Khashayar T, Mahsa T (2010) Biodegradation potential of petroleum hydrocarbons by bacterial diversity in soil. Worl Appl Sci J 8(6):750–755
18. Mbachu AE, Onochie CC, Agu KC, Okafor OI, Awah NS (2014) Hydrocarbon degrading potentials of indigenous bacteria isolated from auto-mechanic workshops at Mgbuka-Nkpor, Nigeria. J Global Biosci 3(1):321–326
19. Thenmozhi R, Nagasathya A, Thajuddin N (2011) Studies on biodegradation of used engine oil by consortium cultures. Adv Env Biol 5(6):1051–1057
20. Namazi BA, Shojaosadati SA, Najafabadi HS (2008) Biodegradation of used engine oil using mixed and isolated cultures. Int J Env Res 2(4):431–440
21. Umanu G, Akpe AR, Omoikhudu AR (2013) Oil degradation assessment of bacteria isolated from used motor oil contaminated soils in OTA, Nigeria. Int J Adv Biol Res 3(4):506–513
22. Adebusoye SA, Ilori MO, Amund OO, Teniola OD, Olatope SO (2006) Microbial degradation of petroleum hydrocarbons in a polluted tropical stream. J Am Sci 2(3):48–57
23. Geetha SJ, Joshi SJ, Kathrotiya S (2013) Isolation and characterization of hydrocarbon degrading bacterial isolate from oil contaminated sites. In: 4th international conference on environmental science and development—ICESD, APCBEE Procedia, Elsevier
24. Santhini K, Myla M, Sajani S, Usharani G (2009) Screening of *Micrococcus* sp. from oil contaminated soil with reference to bioremediation. Bot Res Inter 2(4):248–252
25. Okoh AI (2003) Biodegradation of Bonny light crude oil in soil microcosm by some bacterial strains isolated from crude oil flow stations saver pits in Nigeria. Af J Biotech 2(5):104–108
26. Jesubunmi CO (2014) Isolation of oil—degrading microorganisms in spent engine oil—contaminated soil. J Biol Agric Healthcare 4:191–195

27. Khan JA, Rizvi SHA (2011) Isolation and characterization of micro-organism from oil contaminated sites. Adv Appl Sci Res 2(3):455–460
28. Olajide PO, Ogbeifun LB (2010) Hydrocarbon biodegrading potentials of a *Proteus vulgaris* strain isolated from fish samples. Am J Appl Sci 7(7):922–928
29. Jobson A, Cook FD, Westlake DWS (1972) Microbial utilization of oil. Appl Microbiol 23:1082–1089
30. Mandri T, Lin J (2007) Isolation and characterization of engine oil degrading indigenous microrganisms in Kwazulu-Natal, South Africa. Af J Biotech 6(1):23–27
31. Kumari N, Vashishtha A, Saini P, Menghani E (2013) Isolation, identification and characterization of oil degrading bacteria isolated from the contaminated sites of Barmer, Rajasthan. Int J Biotech Bioeng Res 4(5):429–436
32. Sunita VJ, Dolly PR, Bateja S, Vivek UN (2013) Isolation and screening for hydrocarbon utilizing bacteria (HUB) from petroleum samples. Int J Curr Appl Sci 2(4):48–60
33. Ismail HY, Ijah UJJ, Riskuwa ML, Allamin II (2014) Biodegradation of spent engine oil by bacteria isolated from the rhizosphere of legumes grown in contaminated soil. Int J Env 3 (2):63–75
34. Hamzah A, Rabu A, Farzarul R, Yussoff NA (2010) Isolation and characterization of bacteria degrading Sumandak and South Angsi oils. Sains Malaysiana 39(2):161–168
35. Olawale AK, Akintobi, OA (2011) Biodegradation of glyphosate pesticide by bacteria isolated from agricultural soil, vol 3(1), pp 124–128. http://www.sciencepub.net
36. Barragan-Huerta BE, Costa-Perez C, Peralta-Cruz J, Barrera-Cortes J, Rodriguez-Vazquez Esparza-Garcia F (2010) Biodegradation of organochlorine pesticides by bacteria grown in microniches of the porous structure of green bean coffee. Int Biodeter Biodegr 59:239–244
37. Ortiz-Hernández ML, Sanchez-Selinas E (2010) Biodegradation of the organophosphate pesticide tetrachlorvinphos by bacteria isolated from agricultural soils in Mexico. Rev Int Contam Ambient 26(1):27–38
38. Hindumathy CK, Gayathri V (2013) Effect of pesticide (Chlorpyrifos) on soil microbial flora and pesticide degradation by strains isolated from contaminated soil. J Biodegr Biorem 4(2):1–6
39. Akilandeswari K, Sona V (2013) Efficiency of *Staphylococcus aureus* in the degradation of an organophosphorous pesticide, malathion. J Pharm Sci Innov 2(6):13–21
40. El-Bestawy E, Sabir J, Mansy AH, Zabermawi N (2014) Comparision of the efficiency of different bioremediation technologies of Atrazine-contaminated soils. J Biorem Biodegr 5:237–243
41. Feuntes MS, Benimeli CS, Cuozza SA, Saezl JM, Amorosa MJ (2010) Microorganisms capable to degrade organochlorine pesticides. Curr Res Tech Educ Top Appl Microbiol Microbiol Biotech 1255–1264

Expedient Synthesis of Diverse Spirooxindoles via Multicomponent Approach in Presence of Green Catalyst

Ankita Chaudhary, Pooja Saluja and Garima Khanna

Abstract An efficient, convenient and environmentally benign procedure for the construction of various bioactive spirooxindoles has been developed by condensation reactions of isatins, malononitrile and α-methylene carbonyl compounds/enols in the presence of starch solution as expedient, eco-friendly and biodegradable catalyst at 60 °C. The prominent features of the above protocol are short reaction time, high atom economy, simple work-up, cost-effectiveness, avoidance of toxic chemicals.

Keywords Spirooxindoles · Green synthesis · Multicomponent synthesis Starch

1 Introduction

Spirooxindole based compounds possess wide array of activities such as antitumuor [1], antimicrobial [2, 3], antitubercular [4], antimycobacterial [5], antiproliferative properties [6]. The unique structure of spirooxindoles and their highly pronounced biological activity has attracted the interest among various researchers [7–9].

Moreover design and development of methods to access biologically relevant complex molecules has become increasingly popular at the forefront of contemporary organic synthesis. Multicomponent reactions (MCRs), provide one of the most dominant platforms for the sustainable synthesis of polyfunctionalized heterocyclic compounds owing to their atom economy, operational simplicity, environmental friendliness and green chemistry characteristics [10–13].

A. Chaudhary (✉)
Department of Chemistry, Maitreyi College, Chanakyapuri, New Delhi 110021, India
e-mail: ankitamaitreyi1@gmail.com; achaudhary@maitreyi.du.ac.in

P. Saluja · G. Khanna
Department of Chemistry, University of Delhi, New Delhi 110007, India

V. S. Parmar et al. (eds.), *Green Chemistry in Environmental Sustainability and Chemical Education*, https://doi.org/10.1007/978-981-10-8390-7_2

As a part of our research interest toward the development of environmentally benevolent, efficient and economically viable protocol for the synthesis of heterocycles [14–20], we envisaged on the synthesis of spiropyran annulated heterocycles through one pot condensation of isatins with malononitrile and α-methylene carbonyl compounds/enols. Starch solution which is easily available, inexpensive, biodegradable as well as non-toxic in nature was used as catalyst for the synthesis of aforementioned heterocycles.

2 Results

In the present work, a novel, proficient and green protocol for the synthesis of spirooxindoles has been described via one pot condensation reactions of isatins, malononitrile and α-methylene carbonyl compounds/enols namely 5,5-dimethylcyclohexane-1,3-dione (dimedone), cyclohexane-1,3-dione, 4-hydroxycoumarin, 2,4-dihydro-5-methyl-pyrazol-3-one in presence of starch solution at 60 °C.

In order to optimise the reaction conditions isatin, malononitrile and cyclohexane-1,3-dione were selected as model substrates. Various reactions of isatin (1.0 mmol) with malononitrile (1.5 mmol) and cyclohexane-1,3-dione (1.0 mmol) were attempted in presence of varying amount of starch solution which serve as dual role of catalyst as well as reaction media at different temperature. The impact of different amounts of catalyst load as well as reaction temperature on the yield of desired product i.e. 2-amino-2′,5-dioxo-5,6,7,8-tetrahydrospiro[chromene-4,3′-indoline]-3-carbonitrile (**Ia**) and reaction time is shown in Table 1. The best result was obtained using 5.0 mL starch solution at 60 °C. The reaction was complete in 10 min. and gave 93% of **Ia**.

Table 1 Optimization of reaction conditions for the formation of **Ia**

Entry	Starch solution (mL)	Temperature (°C)	Time (min)	Yield (%)
1	5.0	r.t.	60	55
2	5.0	50	60	75
3	**5.0**	**60**	**10**	**93**
4	5.0	70	10	93
5	4.0	60	25	70
6	6.0	60	10	93
7	8.0	60	10	93

[1]Reactions were carried out in 1.0 mmol scale with 1:1.5:1 ratio of isatin, malononitrile and 1,3-cyclohexane-1,3-dione in starch solution

Using these optimised reaction conditions, reactions of isatin, malononitrile with 3,3-dimethylcyclohexane-1,3-dione/4-hydroxycoumarin/2,4-dihydro-5-methyl-pyrazol-3-one were also performed. All the reactions underwent completion in 10–15 min and afforded the corresponding spirooxindoles (**Ib-d**).

The scope of the above condensation reaction was also examined by using 5-bromoisatin for reactions with malononitrile and 1,3-dicarbonyl compounds (dimedone, cyclohexane-1,3-dione) under otherwise identical conditions and 2-amino-5′-bromo-7,7-dimethyl-2′,5-dioxo-5,6,7,8-tetrahydrospiro[chromene-4,3′-indoline]-3-carbonitrile (**Ie**), 2-amino-5′-bromo-2′,5-dioxo-5,6,7,8-tetrahydrospiro [chromene-4,3′-indoline]-3-carbonitrile (**If**) were obtained respectively. All these results have been shown in Table 2 and represented in Eq. 1.

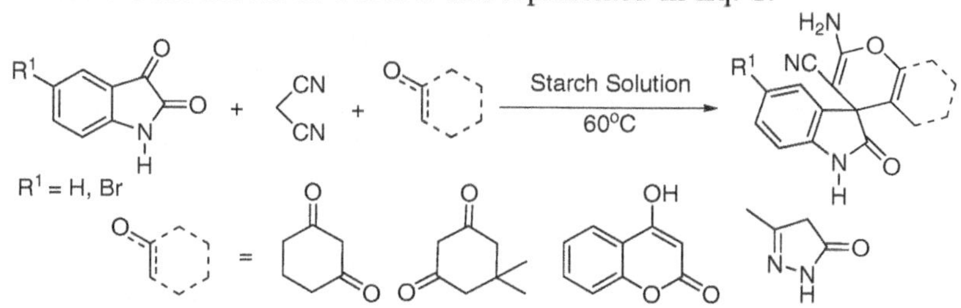

(1)

Table 2 Synthesis of spirooxindoles (I) via reaction of isatins, malononitrile and α-methylene carbonyl compounds/enols in starch solution at 60 °C

Entry	Isatin	α-methylene carbonyl compounds/enols	Product (I)	Time (min)	Yield (%)
1.	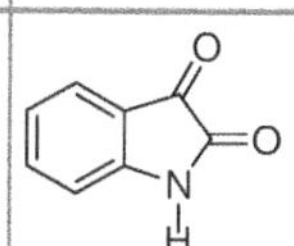	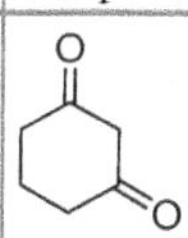	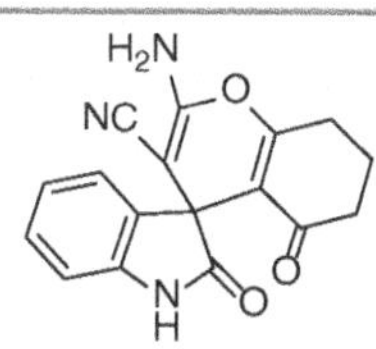 **Ia**	10	93
2.	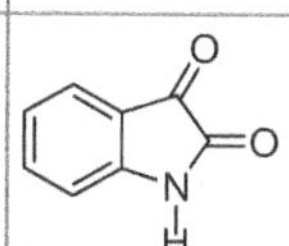	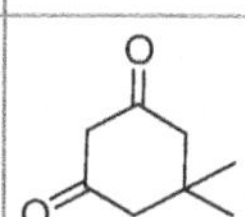	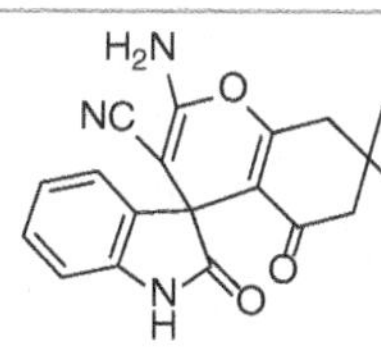 **Ib**	10	94
3.	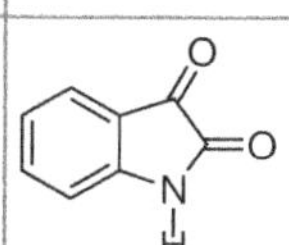	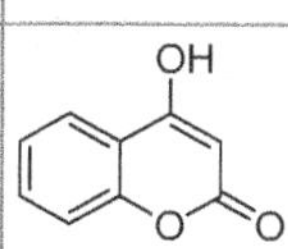	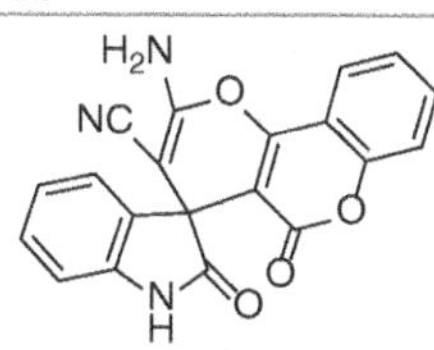 **Ic**	10	95

(continued)

Table 2 (continued)

Entry	Isatin	α-methylene carbonyl compounds/enols	Product (I)	Time (min)	Yield (%)
4.	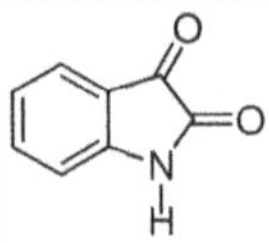	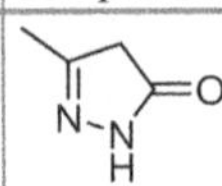	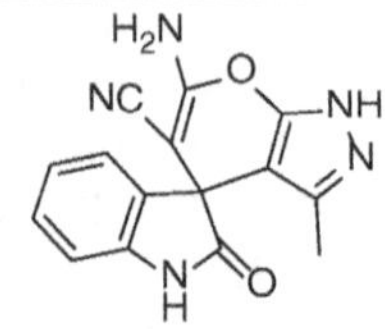 **Id**	15	92
5.	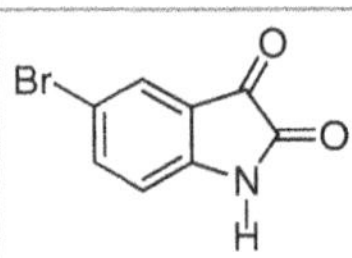	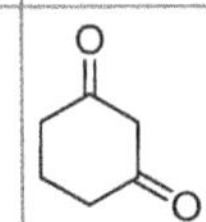	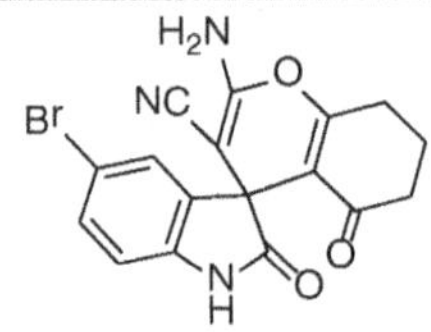 **Ie**	10	91
6.	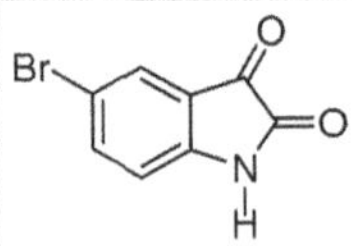	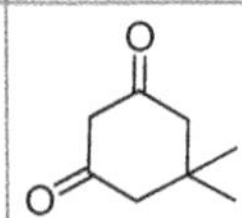	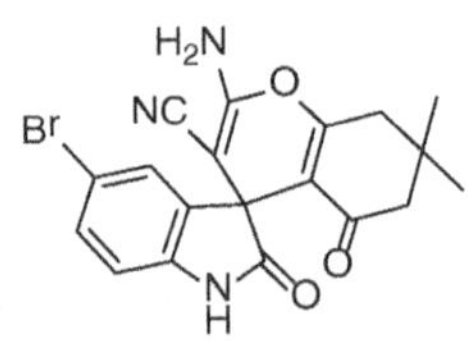 **If**	10	90

[1]Reaction conditions: Isatin/5-bromoisatin (1.0 mmol), 3,3-dimethylcyclohexane-1,3-dione/cyclohexane-1,3-dione/4-hydroxycoumarin/2,4-dihydro-5-methyl-pyrazol-3-one (1.0 mmol) and malononitrile (1.5 mmol) in presence of starch solution (5.0 mL) at 60 °C

3 Discussion

The proposed mechanism for the synthesis of spirooxindoles is shown in Fig. 1. The starch is expected to form micelle-like structure, which is capable of holding the molecules and thereby catalysing the reaction. Initially, Knoevenagel condensation of isatin with malononitrile takes place to give an intermediate. The intermediate so formed underwent Michael addition with α-methylene carbonyl compounds/enol to furnish an intermediate. The hydroxyl group of this intermediate underwent cycloaddition to cyano group to yield **I.**

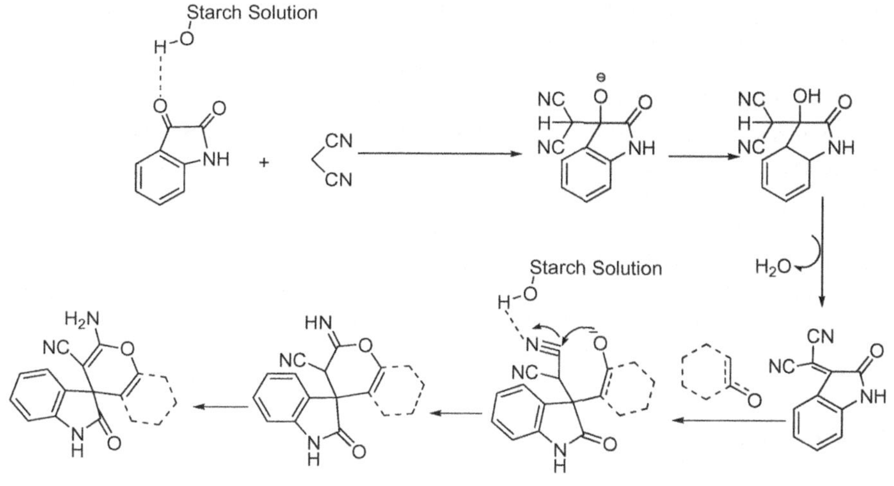

Fig. 1 Mechanism for the formation of spirooxindoles

4 Materials and Methods

All the melting points were measured using Buchi melting point 545 apparatus and are uncorrected. IR spectra were recorded on a Perkin Elmer FTIR spectrophotometer using KBr pellets. The ^{1}H NMR spectra were recorded on Jeol JNM ECX-400P (at 400 MHz) with DMSO-d_6 as solvent and using TMS as internal reference. Thin Layer Chromatography (TLC) was performed on precoated silica gel plates (Merck).

General procedure for preparation of starch solution: The starch solution employed for carrying out synthesis of spirooxindoles was prepared by stirring a mixture of solid starch (1.5 g) in water (15.0 mL) at 25 °C for 30 min. After 30 min. the solution was filtered and the filtrate was used for synthesis of **I**.

Procedure for synthesis of spirooxindoles: To a 50 mL round-bottomed flask isatins (1.0 mmol), malononitrile (1.5 mmol), α-methylene carbonyl compounds/enols (1.0 mmol) and starch solution (5.0 mL) were added. The contents were stirred vigorously at 60 °C for the appropriate times as mentioned in Table 2. After completion of reaction (monitored by TLC), the content was cooled to room temperature. The precipitate so obtained was filtered, washed with water and subsequently with ethanol. All the synthesised products were known compounds and were characterized by FT-IR, ^{1}H NMR and comparison of their melting points with known compounds [21–24].

Spectral Data of Representative Spirooxindoles:

2-Amino-7,7-dimethyl-2',5-dioxo-5,6,7,8-tetrahydrospiro[chromene-4,3′-indoline]-3-carbonitrile (Ib): White solid; Yield = 94%; M.pt.: 300 °C IR (KBr) υ_{max}/cm^{-1} = 3377, 3314, 3145, 2960, 2192, 1722, 1682, 1656, 1471, 1348,

1327, 1223; ^{1}H NMR (400 MHz, DMSO-d_6) δ: 10.32 (1H, s, N$\underline{H}$), 7.15 (2H, s, N$\underline{H}_2$), 7.09-7.04 (1H, m, Ar-$\underline{H}$), 6.91-6.80 (2H, m, Ar-$\underline{H}$), 6.73-6.71 (1H, m, Ar-$\underline{H}$), 2.44-2.43 (2H, m, C$\underline{H}_2$), 2.10 and 2.08 (AB system, 2H, J = 16 Hz, C$\underline{H_aH_b}$C (CH$_3$)$_2$), 0.96 (3H, s, CC$\underline{H}_3$), 0.93 (3H, s, CC$\underline{H}_3$).

2′-Amino-2,5′-dioxo-5′*H*-spiro[indoline-3,4′-pyrano[3,2-*c*]chromene]-3′-carbonitrile (Ic): White solid; Yield = 95%; M.pt.: 294–296 °C; IR (KBr) υ_{max}/cm^{-1} = 3350, 3297, 3196, 2955, 2206, 1736, 1673, 1604, 1541, 1471, 1359, 1219, 1169; ^{1}H NMR (400 MHz, DMSO-d$_6$) δ: 10.69 (1H, s, NH), 7.94 (1H, d, Ar-$\underline{H}$, J = 7.8 Hz), 7.77 (1H, t, Ar-$\underline{H}$, J = 7.7 Hz), 7.66 (2H, s, N$\underline{H}_2$), 7.56 (1H, t, Ar- $\underline{H}$, J = 7.6 Hz), 7.49 (1H, d, Ar-$\underline{H}$, J = 8.4 Hz), 7.22 (2H, t, Ar-$\underline{H}$, J = 7.6 Hz), 6.93 (1H, t, Ar-$\underline{H}$, J = 7.6 Hz), 6.86 (1H, d, Ar-$\underline{H}$, J = 7.9 Hz).

5 Conclusion

In this work, we report a rapid and green synthesis of spirooxindoles via multi-component approach in the presence of starch solution. The benefits of this novel environmentally benign protocol are excellent yields, short reaction time, ease of product isolation and purification, use of environment-friendly catalyst.

Acknowledgements P.S. and G.K. are thankful to UGC and, CSIR, New Delhi respectively, for the grant of junior and senior research fellowship.

References

1. Girgis AS (2009) Regioselective synthesis of dispiro[1H-indene-2,3′-pyrrolidine-2′,3″-[3H] indole]-1,2″(1″H)-diones of potential anti-tumor properties. Eur J Med Chem 44:91–100. https://doi.org/10.1016/j.ejmech.2008.03.013
2. Thangamani A (2010) Regiospecific synthesis and biological evaluation of spirooxindolopy-rrolizidines via [3 + 2] cycloaddition of azomethine ylide. Eur J Med Chem 45:6120–6126. https://doi.org/10.1016/j.ejmech.2010.09.051
3. Bhaskar G, Arun Y, Balachandran C, Saikumar C, Perumal PT (2012) Synthesis of novel spirooxindole derivatives by one pot multicomponent reaction and their antimicrobial activity. Eur J Med Chem 51:79–91. https://doi.org/10.1016/j.ejmech.2012.02.024
4. Rouatbi F, Askri M, Nana F, Kirsch G, Sriram D, Yogeeswari P (2016) Synthesis of new spirooxindole derivatives through 1,3-dipolar cycloaddition of azomethine ylides and their antitubercular activity. Tetrahedron Lett 57:163–167. https://doi.org/10.1016/j.tetlet.2015.11.056
5. Maheswari SU, Balamurugan K, Perumal S, Yogeeswari P, Sriram D (2010) A facile 1,3-dipolar cycloaddition of azomethine ylides to 2-arylidene-1,3-indanediones: synthesis of dispiro-oxindolylpyrrolothiazoles and their antimycobacterial evaluation. Bioorg Med Chem Lett 20:7278–7282. https://doi.org/10.1016/j.bmcl.2010.10.080
6. Yu B, Sun X-N, Shi X-J, Qi P-P, Zheng Y-C, Yu D-Q, Liu HM (2015) Efficient synthesis of novel antiproliferative steroidal spirooxindoles via the [3 + 2] cycloaddition reactions of azomethine ylides. Steroids 102:92–100. https://doi.org/10.1016/j.steroids.2015.08.003

7. Santos MMM (2014) Recent advances in the synthesis of biologically active spirooxindoles. Tetrahedron 70:9735–9757. https://doi.org/10.1016/j.tet.2014.08.005
8. Xia M, Ma R-Z (2014) Recent progress on routes to spirooxindole systems derived from isatin. J Hetereocyclc Chem 51:539–554. https://doi.org/10.1002/jhet.1114
9. Ziarani GM, Moradi R, Lashgari N (2016) Synthesis of spiro-fused heterocyclic scaffolds through multicomponent reactions involving isatin. ARKIVOC (i):1–81. https://doi.org/10.3998/ark.5550190.p009.385
10. Ugi I (2001) Recent progress in the chemistry of multicomponent reactions. Pure Appl Chem 73:187–191. https://doi.org/10.1351/pac200173010187s
11. Tietze LF, Bsasche C, Gericke KM (2006) Domino reactions in organic synthesis. Wiley, New York (2006)
12. Tietze LF (1996) Domino reactions in organic synthesis. Chem Rev 96:115–136. https://doi.org/10.1021/cr950027e
13. Ramon DJ, Yus M (2005) Asymmetric multicomponent reactions (AMCRs): the new frontier. Angew Chem Int Ed 44:1602–1634. https://doi.org/10.1002/anie.200460548
14. Khurana JM, Chaudhary A, Lumb A, Nand B (2012) An expedient four-component domino protocol for the synthesis of novel benzo[a]phenazine annulated heterocycles and their photophysical studies. Green Chem 14:2321. https://doi.org/10.1039/C2GC35644A
15. Khurana JM, Chaudhary A, Nand B, Lumb A (2012) Aqua mediated indium(III) chloride catalyzed synthesis of fused pyrimidines and pyrazoles. Tetrahedron Lett 53:3018–3022. https://doi.org/10.1016/j.tetlet.2012.04.001
16. Saluja P, Chaudhary A, Khurana JM (2014) Synthesis of novel fluorescent benzo[a]pyrano[2,3-c]phenazine and benzo[a]chromeno[2,3-c]phenazine derivatives via facile four-component domino protocol. Tetrahedron Lett 55:3431–3335. https://doi.org/10.1016/j.tetlet.2014.04.072
17. Khanna G, Chaudhary A, Khurana JM (2014) An efficient catalyst-free synthesis of novel benzo[a][1, 3]oxazino[6,5-c]phenazine derivatives via one pot four-component domino protocol in water. Tetrahedron Lett 55:6652–6654. https://doi.org/10.1016/j.tetlet.2014.10.067
18. Khurana JM, Lumb A, Chaudhary A, Nand B (2013) Synthesis and in vitro evaluation of antioxidant activity of diverse naphthopyranopyrimidines, diazaanthra[2,3-d][1, 3]dioxole-7,9-dione and tetrahydrobenzo[a]xanthen-11-ones. RSC Adv 3:1844–1854. https://doi.org/10.1039/C2RA22406B
19. Rajeshwari M, Khanna G, Chaudhary A, Khurana JM (2015) Multicomponent Domino Process for the Synthesis of Some Novel Benzo[a]chromenophenazine Fused Ring Systems Using H2SO4, Phosphotungstic Acid, and [NMP]H2PO4. Synth Commun 45:1426–1432. https://doi.org/10.1080/00397911.2015.1024324
20. Chaudhary A, Gaba R, Sharma K, Upadhyay A, Mathur P, Sharma N, Mishra A, Kapoor K (2016) Green synthesis of spirooxindoles via multicomponent approach. In: Proceedings of national conference in chemistry (NCC 2016): "environmental and harmonious development", New Delhi, India, pp 190–192. ISBN: 9789385824012
21. Khurana JM, Yadav S (2012) Highly monodispersed PEG-stabilized Ni Nanoparticles: proficient catalyst for the synthesis of biologically important spiropyrans. Aust J Chem 65:314–319. https://doi.org/10.1071/CH11444
22. Dabiri M, Bahramnejad M, Baghbanzadeh M (2009) Ammonium salt catalyzed multicomponent transformation: simple route to functionalized spirochromenes and spiroacridines. Tetrahedron 65:9443–9447. https://doi.org/10.1016/j.tet.2009.08.070
23. Li Y, Chen H, Shi C, Shi D, Ji S (2010) Efficient one-pot synthesis of spirooxindole derivatives catalyzed by l-proline in aqueous medium. J Comb Chem 12:231–237. https://doi.org/10.1021/cc9001185
24. Guo RY, An ZM, Mo LP, Yang ST, Liu HX, Wang SX, Zhang ZH (2013) Meglumine promoted one-pot, four-component synthesis of pyranopyrazole derivatives. Tetrahedron 69:9931–9938. https://doi.org/10.1016/j.tet.2013.09.082

Role of Companies in Environmental Management Through Corporate Social Responsibility and Sustainability Policy

Deepika Jindal and Indu Jain

Abstract As the world approaches towards rapid industrialization and globalization, the efficient use of resources along with their conservation has become the mantra for the world to move on sustainable development path. While companies are involved in various activities to produce and trade goods and services to maximise their profit and increase shareholder's wealth, they need to simultaneously focus on Environmental Management, Corporate Social Responsibility and Sustainability policy. The responsibility towards society and environment just no longer lies with the Public Sector and Government but also with the Private Sector. As per the Companies Act 2013, it has become compulsory for large companies to spend 2% of their net profits on Corporate Social Responsibility activities. Indian Companies which are actively working in this area include Mahindra and Mahindra, Tata Steel, Reliance Industries, Hero Motors, Infosys and Wipro to name a few. Various activities undertaken by companies for Environment Management under CSR include water conservation, tree plantation, developing green products, using renewable sources of energy and waste management. The paper aims to analyse the initiatives and steps taken by selected Indian Companies under CSR and Sustainability for Environmental Management and Protection. It also focuses on what areas of environment they have covered and the impact of these initiatives. This is done by analysing secondary sources of information. Also it provides suggestions on various other initiatives that can be taken and how these can be effectively implemented for better future of the country.

Keywords Corporate Social Responsibility · Environmental Policy
Sustainability Policy · Strategy

D. Jindal (✉) · I. Jain
Daulat Ram College, 4, Patel Marg, Maurice Nagar, New Delhi 110007, India
e-mail: deepik12@yahoo.co.in

I. Jain
e-mail: indujain.07@gmail.com

V. S. Parmar et al. (eds.), *Green Chemistry in Environmental Sustainability and Chemical Education*, https://doi.org/10.1007/978-981-10-8390-7_3

1 Introduction

The issues of global warming, environmental degradation, scarcity of resources have become utmost important for everyone, be it the government, public companies, private companies and even the common man as a consumer. It requires integrated efforts by all to tackle these issues so as to make earth a better place for present as well as future generations.

The companies are tackling these environmental issues by developing up a robust Environmental Management System. As per, US Environmental Protection Agency "*An Environmental Management System (EMS) is a set of processes and practices that enable an organization to reduce its environmental impacts and increase its operating efficiency*" [1].

In 1996, International Organization for Standardization (ISO) developed ISO 14001 standards which included the Environmental Management Framework that involves the steps of Plan-Do-Check-Act-Rev Methodology. It involves the following steps:

1. Analysing the impact of processes on the natural environment
2. Planning about environmental initiatives and policies which will minimise the negative impact of human activities
3. Executing these initiatives with the inputs of all stakeholders
4. Continuous monitoring them so as to achieve desired outcome
5. Reviewing to develop better environmental practices

The above steps encompass various sub activities which requires devoting quality time and efforts for effective management of the environment. This provides roadmap to corporates on how to develop and effectively implement their environmental policies.

Environmental Management is an integrated approach with the objective of minimising environmental degradation by taking decisions, following processes and undertaking steps which leads to maximum utilisation, minimal wastage and conservation of natural resources. It can also be referred to as managing the interactions between human beings and environment for developing harmonious relations between them.

The companies are establishing this system through Corporate Social Responsibility and Sustainability Policy. As per International Organization for Standardization's **Guidance Standard on Social Responsibility, ISO 26000,** published in 2010, "*Social responsibility is the responsibility of an organisation for the impacts of its decisions and activities on society and the environment, through transparent and ethical behaviour that:*

- Contributes to sustainable development, including the health and the welfare of society
- Takes into account the expectations of stakeholders

- Is in compliance with applicable law and consistent with international norms of behaviour, and
- Is integrated throughout the organization and practised in its relationships [2]."

As per United Nations Industrial Development Organisation, "*Corporate Social Responsibility is a management concept whereby companies integrate social and environmental concerns in their business operations and interactions with their stakeholders. CSR is generally understood as being the way through which a company achieves a balance of economic, environmental and social imperatives ("Triple-Bottom-Line-Approach"), while at the same time addressing the expectations of shareholders and stakeholders*" [3].

The concept of CSR in India trails backs to 1850s when industrialists used to contribute towards religious causes. Later on with independence, Gandhiji came with the concept of Trusteeship, to devote one's wealth towards societal welfare. In 1960s CSR activities were undertaken majorly by the Public Sector undertakings for protection of the environment and labour welfare. But it was around 1990s that the Private Companies started realigning their business strategies and implementing CSR activities in the fields of environment, women and children welfare, and infrastructural development and so on. It was in 2013, CSR had been mandated by law and incorporated in the Companies Act. Section 135 of this act states that "*every company having: - net worth of Rs 500 crores or more, or - turnover of Rs 1000 crores or more, or - net profit of Rs 5 crores or more during any financial year shall constitute a Corporate Social Responsibility Committee of the Board*" [4]. As per this act, it is compulsory for every company having a net worth of Rs. 500 crores or a turnover of Rs. 1,000 crores or net profit of Rs. 5 crores, to spend at least 2% of its average net profit for the immediately preceding three financial years on CSR activities. The corporates have dedicated resources and teams for planning and executing various CSR programs and initiatives. These help the company not only to increase their operational efficiency but increase their goodwill among the various stakeholders of the company.

The most common definition of Sustainability is given by Brundtland Commission of the United Nations. It states that "*Sustainable development is development that meets the needs of the present without compromising the ability of future generations to meet their own needs* [5]." At 2005 World Summit, it was recognized that sustainability is based on three things—economic development, social development and environmental protection. The companies have a significant role to play in this journey of sustainability as they cannot continue to keep on taking from the environment without giving back to it. A balanced relationship between companies and the environment has to be established. The companies are formulating sustainability policy or environmental policy to transform their systems and processes so as to minimise the adverse effects of their activities on ecological system.

Rio+20 outcome document, Future We Want has emphasised the significance of Corporate Sustainability reports. It stresses that a corporate sustainability report by a private company sets up accountability and transparent standards in its reporting and in this way can contribute more towards sustainable development. The large companies are publishing various reports such as Sustainability Report, Business Processing Report and CSR Report to inform their stakeholders about the work done by them for the environment and society.

As a result, the companies are adopting CSR and Sustainability policy in their business pursuits so as to balance their economic objectives with that of environmental management.

2 Objectives

- To study the steps taken by the selected Indian Companies under Corporate Social Responsibility with respect to environment.
- To analyse how these companies act as environmental stewards through their Sustainability Policy.
- To analyse the areas of the environment on which these companies focus and where they do not.
- To provide suggestions on how the companies can more effectively manage the environment for the welfare of the society.

3 Methodology

The study is an exploratory research. It uses secondary sources such as company's website, annual reports, CSR reports and Sustainability reports and reports by organisations such as United Nations Industrial Development Organization (UNIDO), PricewaterhouseCoopers (PwC) and World Bank. The selection of the companies for analysis has been done on the basis, that these companies have been ranked in top ten in CSR ranking done by Futurescape and Economic Times. The limitation of the study is that the study is based on the reports of those companies which are the top most companies engaged in CSR activities and Environmental Management.

4 Data Analysis and Interpretation

All the corporates be it a manufacturing company or a service provider, are contributing towards environmental management through various programs, initiatives, campaigns and adopting innovative technologies. Sustainability Policy and

Corporate Social Responsibility are the two ways through which companies are working towards maintaining ecological balance.

The paper analyses the steps taken under Sustainability Policy and Corporate Social Responsibility by five corporates in India—Mahindra and Mahindra, Tata Steel, Infosys, Larsen & Turbo and GAIL towards maintaining the environment.

1. INFOSYS—is a leader in technology services and consulting at a global level and has clients in more than 50 countries.
2. Larsen & Turbo—is a conglomerate, operating at global level and its core areas are technology, engineering, construction, manufacturing and financial services.
3. Tata Steel—is the company of famous Tata Group which is in the business of manufacturing steel products.
4. Mahindra and Mahindra—is a federation of companies dealing in automobiles, finance and provides technological solutions. It is guided by its philosophy of Rise—to enable the people to rise.
5. GAIL—is in the business of processing and distribution of natural gas. It is a company which is owned by the government.

Different aspects of environment management, sustainability policy and CSR initiatives of companies have been analysed under the following heads:

a. CSR spending by Companies
b. Areas of Environment Management covered
c. Awards received by selected companies in areas of Sustainability, Environment and CSR

4.1 CSR Spending by Companies

We have analysed the annual reports of the companies for year 2015–16 to find out how much companies are spending on CSR activities and is it as per Companies Act 2013 norm i.e. they are spending two percent of their net profits or not. Also how much of this they are spending on activities related to environment. This is depicted in Table 1.

As can been seen from Table 1, Tata Steel, Mahindra and Mahindra, and GAIL have spent more than the prescribed amount on the CSR activities while the rest were not able to do it. The CSR spending is done in areas of education, women empowerment, environment and rural development. All the five companies are spending only 1–13% of their total CSR spending on environmental management activities. This percentage is very low, seeing that without environment, companies will not be able to operate in future.

Table 1 Amount spent on CSR activities by companies in 2015–16

	Mahindra and Mahindra	Tata Steel	Larsen & Turbo	GAIL	Infosys
Total CSR spending	Rs. 85.90 crores was spent on CSR activities, which was more than (101.12%) of Rs. 84.95 crores, the required amount to be spent on CSR activities	Rs. 204 crores was spent on CSR activities, which was more than (136%) of Rs. 150 crores the required amount to be spent on CSR activities	Rs. 2.35 crores was spent on CSR activities, which was less than (15.73%) of Rs. 14.94 crores, the required amount to be spent on CSR activities	Rs. 118.64 crores was spent on CSR activities, which was more than (115.93%) of Rs. 102.34 crores, the required amount to be spent on CSR activities	Rs. 202.30 crores was spent on CSR activities, which was less than (79.02%) of Rs. 256.01 crores, the required amount to be spent on CSR activities
CSR spending on environment	10.92 crores (12.71% of total CSR spending) were spent on environment sustainability and environment related projects in rural areas	3.27 crores (1.60% of total CSR spending) were spent on environment sustainability	0.01 crore (0.43% of total CSR spending) were spent on environment sustainability	5.45 crores (4.59% of total CSR spending) was spent on environment, healthcare, sanitation and rural development projects	10.34 crores (5.11% of total CSR spending) were spent on environment sustainability

4.2 *Areas of Environment Management Covered*

The following six areas of environment management have been tapped by these companies with their CSR efforts and Sustainability Policy for having a safe planet.

A. Water Management
B. Green Belt
C. Climatic Changes
D. Green Products and Services
E. Energy Management
F. Waste Management

The programs and initiatives undertaken by the five companies in these areas are given below:

A. Water Management

Initiatives taken by selected companies under water management are analysed in Table 2.

Table 2 Initiatives under water management

Mahindra and Mahindra	Tata Steel	Larsen & Turbo	GAIL	Infosys
1. Building watersheds to conserve water for irrigation	1. Going for zero water discharge efforts by operationalizing waste water recovery, commissioning new treatment plants and specific water consumption to drop to the lowest ever 2. At the raw materials locations a switch over to 100% dry beneficiation is under review	1. Adopted zero water discharge system across 28 of its campuses 2. **It has built over 200 check dams which can store 886 million litres of water in Maharashtra in 2013–14** 3. The importance of conserving water has been propagated in all their factories among employees 4. Have set up rain water harvesting systems 5. **Also involved in infrastructure projects of desalination, water management, sewage networks, canal rehabilitation and industrial water management systems**	It target is to reduce water consumption intensity by 45%, and 5% increase in waste water recycling by 2020 by: 1. Zero discharge at Pata and Viaipur plant 2. Treated effluent water is used for horticulture there by saving fresh water 3. Has set up Rain water harvesting facility and rooftop rain water harvesting facility at Vijaipur plant 4. Ensures no water body is adversely by its operations 5. **Jal Sachya project is launched to conserve water by using sprinklers and trigger type water systems to water horticulture** 6. Reducing water wastage by regularly monitoring and repairing leakages from taps and joints	1. **Reduce water wastage by having waterless urinals, key valves and etc.** 2. Has 149 recharge wells and 25 lakes across their campus to harvest rain water 3. **Installed 274 smart water meters across its campuses to monitor its water consumption** **4. Has reduced its fresh water consumption by 41.41% from 2008 to 2016**

As shown in Table 2, water management includes reducing water consumption, harvesting water and effectively disposing water discharges. Most of these companies have set up water sheds, rain water harvesting systems and zero water discharge systems. But Infosys has undertaken certain initiatives different from others such as setting up waterless urinals, developed water smart meters to monitor the consumption of water in their campuses. L&T has built more than 200 check dams thereby increasing their water storage capacity.

B. Green Belt

The steps taken by selected companies under green belt are depicted in Table 3.

Table 3 Initiatives under Green Belt

Mahindra and Mahindra	Tata Steel	Larsen & Turbo	GAIL	Infosys
1. Project Hariyali to plant million trees 2. Adopted Borgad Hill, Nashik—planted 50,000 saplings 3. Replaced paper packaging with paperboard thereby saving 3000 plus trees for production volume of 10000	1. The work on planting trees on the unused areas is under process and is also involved in developing lawns and gardens 3. **With the objective of protecting and restoring the natural habitat of local variety of plants and fruit trees, it has set up Tata Steel Zoological Park at Jamshedpur, botanical parks at mining locations and other afforestation related activities**	1. **Gives Tree Certificates which reflects that they will plant a sapling on behalf of the guest instead of gifting them with flowers** 2. Project Green Hand is a tie up with NGOs to plant more trees at their construction sites, plants, offices, neighbourhood areas, public areas and Himalayan regions. 3. Published a manual on scientific methods of planting trees 4. **Replanted a 158 year only banyan tree at Kilpauk Medical College** 5. Focus on using green packaging wherever applicable	1. Green Harit Project, developed environmental parks for people living in slum areas, Gas crematorium in Delhi and bio gas plants 2. Planted 10,000+ trees at Vijaipura to maintain the local flora 3. **Has launched Mobile Veterinary Services (MVS) units in the North-East to protect wildlife**	1. Planted around 80621 trees in 2015–16 across their campuses and promotes planting native plants to support local fauna

As analysed from Table 3, tree plantation is the most common initiative undertaken by all the companies under CSR. Besides that the companies are involved in developing greener areas in their factories as well as offices. Tata Steel has developed Zoological parks to protect natural flora and fauna. L&T gives trees certificates to their guests instead of flowers. The certificate states that L&T will plant tree on their behalf. GAIL supports wildlife through their Mobile Veterinary system for protecting the wildlife.

C. Climatic Changes

The selected companies have undertaken various steps with respect to climatic changes that are given in Table 4.

Table 4 Initiatives under climatic changes

Mahindra and Mahindra	Tata Steel	Larsen & Turbo	GAIL	Infosys
1. Hosted carbon pricing workshop to encourage Indian corporates to take action on carbon pricing 2. Continuously monitor their GHG emissions	1. Active member in various committee working towards reducing carbon emission and climate change—Prime Minister's low carbon committee, 'Caring for Climate' initiative of the United Nations Global Compact and United Nations Environment Programme 2. Three plant sites in Thailand operate electric arc furnaces that restrict the emission of carbon dioxide to the environment 3. An environment department has been set up inside the works so as to continuously observe the air and water pollution levels and undertake corrective steps 4. Company has a policy on climate change	1. Have invested in low carbon technologies for building infrastructure. 2. Conducted studies at certain locations on reducing carbon footprints mapping. 3. Working on reducing carbon di oxide emissions of around 16,128 per year through its development wing (L&T IDPL). The project for it is called clean development mechanism (CDM) and is listed with United Nations Framework Convention on Climate Change (UNFCCC) 4. Is committed to the eight missions of the	Target is to reduce GHG emission intensity by 33% by 2020: 1. Undertaking research such as signed MOU with USEPA to carry out studies regarding fugitive methane emissions 2. Capturing landfill gas at Ghazipur which is used to generate electricity through a micro turbine 3. **Converting Carbon di oxide to Syn gas and fixing it using micro algae for production of biofuels** 4. It has installed pollution control devices at its	1. **Aims to be Carbon neutral by 2018** 2. Have reduced their carbon emission by 58% across scope 1 and 2 emissions

(continued)

Table 4 (continued)

Mahindra and Mahindra	Tata Steel	Larsen & Turbo	GAIL	Infosys
	5. **Launched GREENFECTION: The Green School is an initiative on various problems with climate change which includes articles on climate change, interactive rain water harvesting game, a calculator to help estimate your carbon footprint, news from around the world and a lot more**	National Action Plan on Climate Change (NAPCC) instituted by the Government of India 5. **Provides training to its employees on subject of climate change and sustainable development** 6. Participates in various forums, seminars on climate change and sustainability 7. **L&T uses machines which causes less noise and air pollution by developing an integrated management system**	various units for reducing air emissions 5. Fixed and mobile monitoring stations to continuously check air quality 6. In 2015, it has partnered with TERI to create climate change awareness 7. **Celebrated "Climate Change Awareness Week" and a quiz on the same** 8. Presented its contribution to tackle climate change at COP21, Paris as part of a delegation of MoPNG	

Table 4 shows that global warming, changing climate and greenhouse gas emissions is one of concern areas for all industries as they adversely affect the health of human beings. It also causes severe shifts in the environmental processes. It is not just government but companies in different countries are also making efforts to reduce their carbon emissions. All these five companies are making efforts to monitor and reduce their GHG emissions by using better equipment. They have actively organised and participated in various conferences, seminars and workshops on Climatic changes. Tata Steel has launched GREENFECTION, a Green School is an initiative on various problems with climate change which includes articles on climate change, interactive rain water harvesting game, a calculator to help estimate your carbon footprint, news from around the world and a lot more. L&T provides training to its employees on how they can contribute to resolving this issue.

D. Green Products and Services

The initiatives taken by these five companies for providing green products and services are given in Table 5.

From the above Table 5, it emerges that the consumers of today demand green products and services which are eco-friendly in all aspects be its manufacturing, usage or disposable. Mahindra and Mahindra is manufacturing fuel efficient cars, Tata steel is developing eco-friendly high steel, GAIL produces natural gas and polymers and L&T is constructing Mumbai monorail. All these products are eco-friendly. But it is not just manufacturing sector, also service sector is going greener-by developing greener technology such as radiant cooling and developing energy management software. Also, Infosys have total of 14 LEED (Leadership in

Table 5 Initiatives under green products and services

Mahindra and Mahindra	Tata Steel	Larsen & Turbo	GAIL	Infosys
1. Launched various fuel efficient vehicles such as New XUV 5OO, Verito Vibe and GenZe 2.0 with zero emissions	1. **Committed towards designing high steel which is sustainable in nature that is have longer life and requires few resources for its production**	1. Increased green products and service in its portfolio by 12% such as energy management softwares, energy audits from 2012–13 to 2013–14 and this have led to increase in its sale by 14% in same year 2. Developed solar water pumps to minimise electricity usage and thereby helping farmers 3. **Constructed the Mumbai Monorail, an eco-friendly system that saves fuels and carbon emissions** 4. Process of setting up of Syn Gas power plants	1. The pipeline network at GAIL supplies natural gas LPG to industries and households in more than 31 cities, thereby saving transportation fuel and reducing pollution 3. **Produces products such as polymers which are environment friendly and completely recyclable** 4. Creating awareness among consumers about use of plastics as they believe it can be recyclable by having plastic waste management system	1. Implemented various green technologies such as radiant cooling technology in their buildings for energy conservation 2. Planted Beema bamboos in their campuses to reduce carbon footprint and it act as an energy crop for generating charcoal, ethanol through gestification 3. **Have total of 14 LEED (leadership in environment and development design) and 4 GRIHA (green rating for integrated habitat assessment) buildings across their campus**

Environment and Development Design) and 4 GRIHA (Green Rating for Integrated Habitat Assessment) buildings across their campus, which helps in conserving electricity.

E. Energy Management

The selected companies have undertaken various steps with respect to Energy Management which are depicted in the Table 6.

Table 6 Initiatives under energy management

Mahindra and Mahindra	Tata Steel	Larsen & Turbo	GAIL	Infosys
1. Set up various solar power plants, Mahindra Susten is the largest 2. Installed various solar telecom towers 3. Generating wind energy at their Khopol Pant 4. Project Suryodaya—electrified 66 households of villages of Belwa and Bashi and created e-hubs for livelihood generation avenue 5. First company to signed EP 100—it is a drive being done at global level aiming to encourage significant companies to commit themselves to double their energy productivity	1. Uses judiciously finite iron ore, chromite and coal as raw material by implementing innovative ore beneficiation and extraction techniques, resource efficient manufacturing and minimising scrap generation, eco-friendly mining and fast restoration of mined out areas	1. In 2014–15, 11% of their indirect energy mix comprises of renewable energy from solar, wind and biogas 2. **Has more than 19 trained energy auditors** 3. **Have training centres to provide generic training on good electrical practices** 4. Uses energy saving services and energy efficient electrical equipment 5. It has built 187 Mwp solar plants till date and has completed building one of the largest power plant of thermal energy in Asia 6. It continues to tap solar energy in its various plants such as Chennai, Talegaon, Mahape and etc.	It targets to reduce 5% of specific energy in liquid hydrocarbon and polymer products by: 1. Monitoring its energy consumptions 2. Conducting energy audits and taking corrective actions 3. Using renewable energy by installing online solar panels, 3 solar power plants, solar system with battery backup, solar based lightening at its plants 4. Adopting ISO50001:2011 energy management system. 5. Utilises energy efficient products at its facilities such as 3 star electrical equipment, LED based lighting system	1. Solar plant at Hyderabad that can meet their total electricity requirement 2. At their Indian locations their energy consumption reduced by 6.54% from 2016 to 2015 3. Set up the goal to meet all their energy requirements from renewable energy. 4. **One of the first companies to join RE100 platform.** 5. **Have a Centralised Energy Management system to monitor energy consumption and to avoid energy wastage at their campuses through SMS services** 6. Using retrofit programs to conserve energy 7. Partnered with various organisations

(continued)

Table 6 (continued)

Mahindra and Mahindra	Tata Steel	Larsen & Turbo	GAIL	Infosys
		7. **Save electricity by having software for hibernating desktops and having cooling mechanism at data centres** 8. Have replaced metal lamps with energy efficient lightening such as CFL and T5 fittings 9. Set up a Solar training institute with Ministry on New and Renewable Energy	6. **Trades water renewable energy certificates for acquiring wind energy for Tamil Nadu** 7. Construction of energy efficient gas plants	worldwide such as National Renewable Energy Laboratory (US), 3M, and Saint Gobain for conducting research in energy efficiency

As shown in Table 6, Energy Management comprises of reducing energy consumption and opting for alternative sources of energy. All these companies are committed towards reducing their consumption of electricity. Also they are developing techniques to harvest and utilise alternative sources of energy such as hydro energy, solar energy and wind energy. Mahindra and Mahindra has electrified two villages and Tata Steel is using judiciously iron ore and coal by deploying innovative technology. GAIL conducts energy audits and L&T provides training to its employees on good electrical practices and it has set up a Solar Training Institute with Ministry on New and Renewable Energy. Infosys have a Centralised Energy Management system to monitor energy consumption and to avoid energy wastage at their campuses through SMS services.

F. Waste Management

The steps taken by the selected companies for managing waste are given in Table 7.

Table 7 shows that most of the companies have set up bio gas plants to convert kitchen waste into fuels that can be used for cooking. Tata Steel has developed Bio toilet and Bio digester to convert human waste into odourless water and inflammable gas. All of them have effective waste disposable systems so as not to harm nearby communities. Infosys through its Rural Reach Program provides guidance to schools on disposable of e waste and has partnered with a NGO to setup bio gas plants in districts of Karnataka.

Table 7 Initiatives under waste management

Mahindra and Mahindra	Tata Steel	Larsen & Turbo	GAIL	Infosys
1. Setup 'Bioneer', an organic waste convertor in Mahabaleshwar to convert organic waste into compost in just 24 h 2. Try to minimise wastage in their manufacturing process	1. **Developed bio toilet and bio digester to convert human waste into odourless water and inflammable gas**	1. Builds waste management system for its customers 2. Focuses on reducing waste at all it its plants, factories and offices by following effective waste management system—proper disposable of hazardous waste, recycling the waste and regular checking and reporting systems 3. Recycles zinc and steels at its plant and promote use of recycle paper 4. **Solid and liquid resource management project to segregate organic and inorganic waste by providing buckets at 14 panchayats in Chennai**	1. E-waste management by providing training to its officials on disposal techniques and collection system 2. Has well developed systems and uses various methods such as recycling, waste treatment, incineration to dispose the solid, liquid and miscellaneous waste. 3. Hazardous substances are disposed by connecting valves to a highly developed flare system where they are burnt under control conditions 4. **Plans to set up two 'waste plastic to fuel oil' plants**	1. **Through rural reach program it provides guidance to schools on disposable of e waste** 2. Set up bio gas plants and organic waste convertors at some of its campuses to convert food waste into bio gas for kitchens and manure for landscapes. 3. Partnered with SKG Sangha an NGO to build around 7620 bio gas units in Ramanagara district in Karnataka to provide alternative source of fuel for kitchens and thereby reducing pollution

4.3 *Awards Received by Selected Companies in Areas of Sustainability, Environment and CSR*

Table 8 contains the details of the awards won by these five companies in areas of Sustainability, Environment and CSR in year 2015–16.

Table 8 Awards on sustainability and CSR

2015–16				
Mahindra and Mahindra	Tata Steel	Larsen & Turbo	Gail	Infosys
Received Sustainable Plus Platinum Label for Financial Year 2015 from CII-CESD (Centre of Excellence for sustainable Development) • Received 'Excellence in Sustainable Supply Chain Award 2015' in the CII-ITC Sustainability Awards • Ranked first for CSR in 'The Best Companies for CSR' study conducted by The Economic Times (Futurescape and IIM Udaipur) • Received 'Golden Peacock Award for Risk Management' from the Institute of Directors	1. Golden Peacock Global Award for Sustainability in 2015 2. Environment Award, 2015 by JSPCB, for Scientific Water Conservation, Over Burden Management and Afforestation, won by Noamundi Iron Ore Mine 3. In February 2016, Jharia Division won the Global Safety Award (Gold Category) By Energy & Environment Foundation	1. Frost & Sullivan Excellence Award was won by L&T Technology Services Wins 2. L&T Ranked Asia's 2nd most sustainable company in industrial sector	1. In 2015, GAIL scored 98 and was ranked 3rd in CDP India's "Carbon Disclosure Leadership Index (CDLI)"	1. It was awarded the Golden Peacock Award at the 16th London Global Convention on 'Corporate Governance and Sustainability' 2. Was presented with the Innovative Technology of the year award at The Energy Award, London 3. At Federation of Indian Chambers of Commerce and Industry Water Awards was positioned third in the Industrial Water Use efficiency category 4. Was awarded the NDTV Property Award for the Most Environment—Friendly Commercial/ Office Space Category

(continued)

Table 8 (continued)

2015–16				
Mahindra and Mahindra	Tata Steel	Larsen & Turbo	Gail	Infosys
• Continued to be listed on the Dow Jones Sustainability index—2015 under the 'Emerging Market Index' for the consecutive third year with improvement in percentile scores • Has continued to be in the top 10 in the India 200 Carbon Disclosure Leadership Index 2015 • The Mahindra Group Sustainability Report was shortlisted for the finalists of Asia Sustainability Reporting Awards by CSRworks				5. The Solar Energy Global Conference and Awards

The importance of company's responsibility towards the society and environment can be seen from the number of awards being presented in these areas by various international and national organisations. These awards not only motivate the companies to make more efforts in these areas but also creates positive image about the companies in the minds of its various stakeholders.

5 Conclusions

- Under Corporate Social Responsibility, most of the companies are mainly directing their initiatives towards health, education and women and thus less towards the environment.

- With regard to CSR spending on Environment related initiatives, they are spending only 1–13% of the total amount spent on all CSR activities
- With respect to environmental initiatives under CSR, the following is observed:
  - All the five companies are planting more trees across different regions of India and some have even adopted certain villages
  - Very few are working towards wildlife conservation
  - Some of them have set up bio gas plants in village, thereby providing villagers alternative source of energy
  - Employee's campaigns are being organised on issues related to global warming and conserving the environment.
- These companies are working towards environment management primarily by focusing their business strategy towards sustainable development. This includes:
  - Having code of conduct stressing on environment, safety and health
  - Developing production processes and utilising equipments which are eco-friendly in nature
  - Producing green products and services which do not adversely affect the environment
  - Going for e-waste management and recycling the waste to produce biogas and using it as fuel
  - Creating awareness among the employees to judiciously use the natural resources and how they can conserve it
  - The companies are reducing energy consumption in their offices and buildings besides their manufacturing plants.
- Almost all the corporates are undertaking various steps to develop and harness renewable sources of energy especially the solar energy by setting up solar panels.

6 Recommendations

- The companies need to undertake more initiatives under CSR for conservation of environment.
- The CSR initiatives towards environment should not just focus on plantation only but extend to other areas also such as utilising renewable energy and generating alternative sources.
- The companies should help communities to adopt energy saving practices which they are following in their offices such as water and electricity consumption monitor.

- The companies can collaborate:
  - With each other towards Environmental Management.
  - With NGOs for effective implementation of their campaign
  - With their suppliers and vendors to have eco-friendly distribution network
- Training sessions, conferences and seminars should be organised to disseminate information about various environmental issues and how they can be resolved
- They should launch more campaigns to create awareness among the public about the disposable of e-waste.

References

1. United States Environmental Protection Agency. https://www.epa.gov/ems. Accessed 25 Oct 2016
2. The Australian Centre for Corporate Social Responsibility. http://accsr.com.au/what-is-csr/. Accessed 30 Oct 2016
3. United Nations International Development Organization. https://www.unido.org/csr/o72054.html. Accessed 1 Nov 2016
4. Bahl E (2014) An overview of CSR rules under companies act, 2013. http://www.business-standard.com/article/companies/an-overview-of-csr-rules-under-companies-act-2013-114031000385_1.html. Accessed 2 Nov 2016
5. Wikipedia. https://en.wikipedia.org/wiki/Sustainable_development. Accessed 26 Oct 2016

Further Reading

6. World Bank (2002) Definition of emerging economies, global economic prospects and the developing countries. www.sustainability.com/developing-value/definition-emerging.asp. Accessed 29 Oct 2016
7. Samuel J, Saari A (2007) Corporate social responsibility. http://infochangeindia.org/corporatesrlbP.jsp. Accessed 1 Nov 2016
8. Blowfield M, Murray A (2008) Corporate responsibility—a critical introduction. Oxford University Press, New Delhi
9. Crowther D, Aras G (2008) Corporate social responsibility. Ventus Publishing ApS
10. International Financial Corporation. The role of the private sector in natural resource management. A focus on forests. http://siteresources.worldbank.org/INTENVMAT/64199955-1162240805462/21127302/16IFC.pdf. Accessed 1 Nov 2016
11. Arun MG (2015) Green is in for India Inc. http://indiatoday.intoday.in/story/green-business-environment-indian-companies/1/442275.html. Accessed 29 Oct 2016
12. Mahindra and Mahindra Limited. http://www.mahindra.com/resources/pdf/sustainability/Sustainability-Snapshot.pdf. Accessed Oct 2016
13. Tata Steel Limited. http://www.tatasteelindia.com/corporate-citizen/environment/environment-management.asp. Accessed Oct 2016
14. Handbook on Corporate Social Responsibility in India (2013) PricewaterhouseCoopers Private Limited (PwCPL). https://www.pwc.in/assets/pdfs/publications/2013/handbook-on-corporate-social-responsibility-in-india.pdf. Accessed Oct 2016

15. Infosys Limited. https://www.infosys.com/sustainability/Documents/infosys-sustainability-report-2015-16.pdf. Accessed Nov 2016
16. Larsen & Toubro, India. http://www.larsentoubro.com/media/32502/lt-sustainability-report-2015-dec-15.pdf. Accessed Nov 2016
17. GAIL (India) Limited. http://www.gailonline.com/final_site/pdf/SR_14-15_Final.pdf. Accessed Nov 2016

The Use of Spectroanalytical Techniques in Detecting the Heavy Metal and Salinity in *Solanum tuberosum* (cv. Kufri Bahar) During Indoor Study

Govinda, Asha Sharma and Sandeep Singh

Abstract The vegetable crops are getting most affected by exposure to rising concentrations of heavy metals and salinity in soil all over the world. Their accumulation in such plants can be analyzed by sophisticated techniques for further studies. Atomic absorption spectroscopy (AAS) is very popular spectroanalytical technique used to quantify nearly seventy different elements in solution or directly in solid samples. The flame photometry is used to estimate the Na (sodium) in plants as marker of NaCl level. The indoor growth of the *Solanum tuberosum* was done to check the effects of Cd and salinity in roots and shoots. The nutrients are provided into Hoagland's solution. The regimes of cadmium, sodium and combination of both stresses were chosen. The cadmium accumulation increased all over the plant but highest level in root was at 2 mM Cd given without NaCl. Root Na increased with increase in external NaCl irrespective of Cd. But, shoot Na levels behaved differently because Na increase peaked at 50 mM NaCl when given individually but increase was continuous when NaCl was given in presence of cadmium. AAS and flame photometry aided in getting a clear picture of Cd and Na accumulation in potato plants.

Keywords Cadmium · NaCl · Potato · Atomic absorption spectroscopy and flame photometry

Govinda · S. Singh (✉)
Department of Biochemistry, MD University, Rohtak, Haryana, India
e-mail: Sandeep_rtk@yahoo.com

Govinda
e-mail: govind22053@gmail.com

A. Sharma
Department of Botany, MD University, Rohtak, Haryana, India

V. S. Parmar et al. (eds.), *Green Chemistry in Environmental Sustainability and Chemical Education*, https://doi.org/10.1007/978-981-10-8390-7_4

1 Introduction

Salt stress is a major environmental factor limiting the productivity of crops by negative affects on the metabolism of plants and causing significant alterations in different biochemical and molecular processes [1]. Salt stress triggers a number of interacting events like inhibition of photosynthetic enzymes activity, membrane damage and the disruption of membrane structures [2]. Other effects are induction of osmotic stress, oxidative stress and protein denaturation in plants which lead to cellular adaptive responses and accumulation of compatible organic solutes like soluble carbohydrates, amino acids, proline, betaines etc. [3, 4]. This organic solute accumulation can alleviate stress by removal of free radicals, and stabilization of macromolecules and organelles, such as proteins, protein complexes and membranes [5]. Plants fight oxidative stress by various antioxidant compounds and detoxifying enzymes [6]. The activity and expression levels of the genes encoding detoxifying enzymes are also probably enhanced by reactive oxygen species (ROS) under abiotic stresses.

Cd also impedes plant growth by interference in photosynthesis, respiration and water relations apart from uptake and translocation of mineral nutrients, which all can significantly alter the normal plant growth [7]. Different Plants respond to cadmium in molecularly regulated process called metal homeostasis. This also includes regulation of the metal induced ROS signaling pathway [8]. Improper management of waste and application of chemicals containing Cd, especially in developing countries, have severely increased cadmium contamination in arable soils and surface water [9]. Cadmium, a non-essential element, is released into the environment by power stations, heating systems, metal-working industries or urban traffic. The soil solutions containing 1 mM or more Cd are considered to be highly polluted [10]. Plant Cd concentrations are between 0.05 and 0.2 ppm [11], but can be much higher in plants growing in contaminated soils. First step in studies of abiotic stresses on plant system is assessment of abiotic toxicant accumulation in plant body. Thus present study involved analysis of sodium (Na) and cadmium (Cd) content as hallmarks of NaCl and Cd stresses on the chosen plant *Solanum tuberosum*.

2 Results

The levels of Na in roots increased with increase in external NaCl whether Cd was present or not (Fig. 1). But, shoot Na levels behaved differently because Na increase peaked at 50 mM NaCl when given individually but increase was continuous when NaCl was given in presence of cadmium. The increase in shoot Na was reported earlier in 2 potato cultivars in 0–120 mM NaCl [12]. Recently, Jaarsma et al. (2013) also reported increase in shoot Na levels of two sensitive and

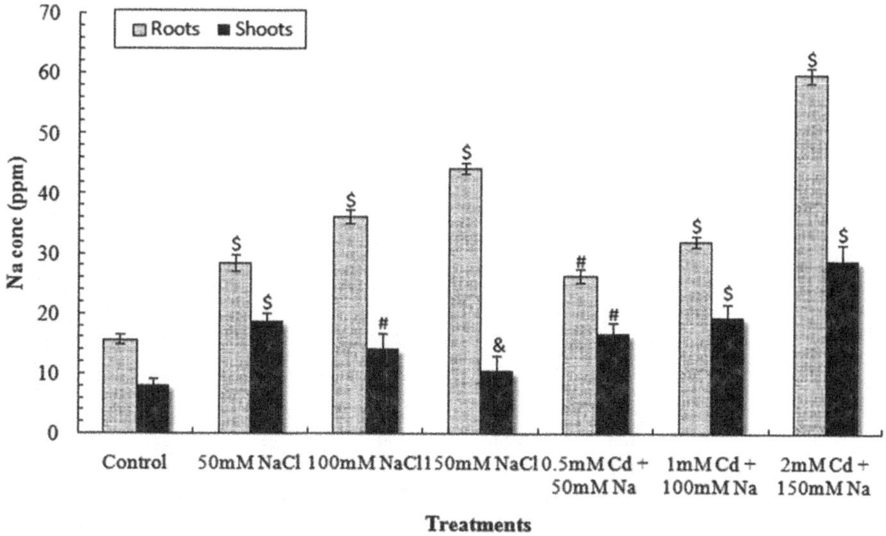

Fig. 1 The graph representing the accumulation of the sodium by roots and shoots of the kufi bahar cultivar of *Solanum tuberosum* in various stresses ($p < 0.05$ for &; $p < 0.01$ for #; $p < 0.001$ for $; change non-significant if none of these symbols shown)

two tolerant potato cultivars grown in 0/60/180 mM NaCl for one week was reported [13].

The root cadmium level increased with increase in external cadmium when given individually or with NaCl but highest level was at 2 mM Cd given without NaCl (Fig. 2). The shoot cadmium also increased whether NaCl given or not. The retention of cadmium was more by the roots than shoots. Increase in Cd was also reported in leaves and roots of 6-week old hybrid poplar cultivated in 0/10/100 μM Cd for 6 more weeks [14], for 14-day old two maize cultivars treated with 0/0.3/0.6/ 0.9 mM $Cd(NO_3)_2$ for 8 days [15], for *Cucumis sativus* grown in 0/10/50/100/ 500 μM Cd for 5/10/15/20 days [16] and for 30-day old tomato plants treated with 0/2.5/5/10/20 μM Cd for 30 more days [17]. Cd and Na estimation revealed some usual outcomes. Both metals generally increased in both tissues with increase in concentration in growth medium. But, Na decreased in shoot from 50 mM to higher levels but increased in presence of Cd. Sodium content at 150 mM NaCl was higher in presence of 2 mM Cd otherwise there was no difference in Na with or without Cd. The content of cadmium also was similar with or without Na.

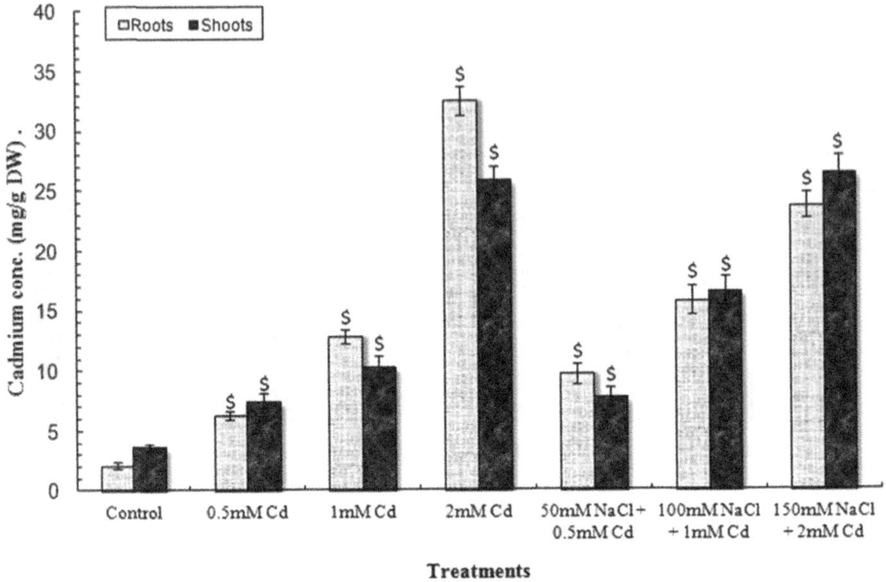

Fig. 2 The graph representing the gathering of the cadmium by roots and shoots of the kufi bahar cultivar of *Solanum tuberosum* in various stresses ($p < 0.05$ for &; $p < 0.01$ for #; $p < 0.001$ for $; change non-significant if none of these symbols shown)

3 Discussion

Plants are negatively affected by some heavy metals also and cadmium is an important heavy metal in this group. Improper management of waste and application of chemicals containing Cd, especially in developing countries has lead to cadmium becoming environmentally important toxic metal. It is one of the most readily absorbed and most rapidly translocated heavy metal in plants and so exerts strong toxicity even at relatively low concentrations. Plant Cd concentrations are between 0.05–0.2 ppm, but can be much higher in plants growing in contaminated soils. Hence present study was undertaken to study accumulation of Cd and NaCl in plant in these stresses. Cd and Na estimation revealed some usual outcomes. Both metals generally increased in both tissues with increase in concentration in growth medium. But, Na decreased in shoot from 50 mM to higher levels but increased in presence of Cd. Sodium content at 150 mM NaCl was higher in presence of 2 mM Cd otherwise there was no difference in Na with or without Cd. The content of cadmium also was similar with or without Na. However, this accumulation can further analyzed in greater detail by carrying out subcellular studies.

4 Materials and Methods

4.1 Materials

For the present study, Kufri bahar cultivar of potato (*Solanum tuberosum* L.) was procured from cold store prescribed by CPRI, Kufri (Shimla). All chemicals were of AR grade. Sample preparations were done in main lab. Cd estimations was done by AA spectrophotometer (SENSAA, Make—GBC Australia) at GJU (Hisar) and Na estimations was by flame photometer (Model—CL 378, Elico India) at GJU (Hisar).

4.2 Plant Cultivation

First soil samples were collected from agricultural fields and their pH/metal contents/NPK contents were tested at Krishi Vigyan Kendra (HAU), Rohtak and very low contents of Na and Cd were confirmed. The soil samples were thoroughly mixed with natural manure and loaded into the pots. The pots were lighted 4–5 h daily by fluorescent lights in the range 400–700 nm sufficient for photosynthetic activities. The pots were irrigated by normal water for 2 days for moisture maintenance. Then 5 potato tubers were planted per pot for total of 10 pots for control and 9 treatments. The treatments were 0.5 mM/1 mM/2 mM Cd, 50 mM/100 mM/150 mM NaCl and their corresponding combinations. These were all prepared in Hoagland's nutrient solution [18]. The treatments were continued for 28 days and plant shoots and roots were harvested at each week except 1st one when growth was insufficient. So, results were reported for 2nd, 3rd and 4th weeks.

4.3 Na and Cd Estimation

The harvested root and shoot portions were oven dried at 80 °C for 24 h and weighed for dry mass. The tissues were digested in digestion mixture of HNO_3 and $HClO_4$ in 9:1 ratio in 0.5:15 digestion mix (w/v). The mixtures were kept overnight and then heated on hot plate till solution became clear. Then, they were collected into containers for Na/Cd estimations.

5 Conclusions

Both Na and Cd increased in both tissues with increase in growth medium. Sodium decreased in shoot from 50 mM to higher levels but increased in presence of Cd. The change in sodium content are largely independent of cadmium while content of cadmium also was similar with or without Na.

Conflicts of Interest The authors declare no conflict of interest.

References

1. Allakhverdiev SI, Sakamoto A et al (2000) Ionic and osmotic effects of NaCl-induced inactivation of photosystems I and II in *Synechococcus* sp. Plant Physiol 123(3):1047–1056
2. Meloni DA, Oliva MA et al (2003) Photosynthesis and activity of superoxide dismutase, peroxidase and glutathione reductase in cotton under salt stress. Environ Exp Bot 49:69–76
3. Hasegawa PM, Bressan RA et al (2000) Plant cellular and molecular responses to high salinity. Ann Rev Plant Physiol Plant Mol Biol 51:463–499
4. Munns R (2002) Comparative physiology of salt and water stress. Plant Cell Environ 28:239–250
5. Bray EA, Bailey-Serres J et al (2000) Responses to abiotic stress. In: Buchanan B, Gruissem W, Jones R (eds) Biochemistry and molecular biology of plants. American Society of Plant Physiology, Rockville MD, USA, pp 1158–1203
6. Ashraf M, Akram NA et al (2010) The physiological, biochemical and molecular roles of brassinosteroids and salicylic acid in plant processes and salt tolerance. Crit Rev Plant Sci 29:162–190
7. Gill SS, Tuteja N (2011) Cadmium stress tolerance in crop plants: probing the role of sulfur. Plant Signal Behav 6:215–222
8. Lin YF, Aarts MGM (2012) Molecular mechanism of zinc and cadmium stress response in plants. Cell Mol Life Sci 69:3187–3206
9. Mühling KH, Läuchli A (2003) Interaction of NaCl and Cd stress on compartmentation pattern of cations, antioxidant enzymes and proteins in leaves of two wheat genotypes differing in salt tolerance. Plant Soil 253:219–231
10. Sanita di Toppi L, Gabrielli R (1999) Response to cadmium in higher plants. Environ Exp Bot 41:105–130
11. El-Bassam N (1978) Superelement: Nahrstoffe and Gift Zugleich. Kali-Bnefe (Buntehof) 14:255–274
12. Aghaei K, Ehsanpour AA et al (2009) Potato responds to salt stress by increased activity of antioxidant enzymes. J Integ Plant Biol 51:1095–1103
13. Jaarsma R, de Vries RSM et al (2013) Effect of salt stress on growth, Na^+ accumulation and proline metabolism in potato (*solanum tuberosum*) cultivars. PLoS ONE 8(3):e60183
14. Nikolic N, Kojic D et al (2008) Responses of hybrid poplar to cadmium stress: photosynthetic characteristics, cadmium and proline accumulation, and antioxidant enzyme activity. Ac Biol Cracov Ser Bot 50:95–103
15. Ekmekci Y, Tanyolac D et al (2008) Effects of cadmium on antioxidant enzyme and photosynthetic activities in leaves of two maize cultivars. J Plant Physiol 165:600–611
16. Sun S, Li M et al (2015) Cadmium effects on mineral accumulation, antioxidant defence system and gas exchange in cucumber. Zemdirbyste-Agric 102:193–200

17. Khavari-Nejad RA, Najafi F et al (2013) The effects of GA_3 application on growth, lipid peroxidation, antioxidant enzymes activities, and sugar levels of cadmium stressed tomato (*Lycopersicon esculentum mill. Cv. CH*) plants. Rom J Biol Plant Biol 58:51–60
18. Dguimi HM, Debouba M et al (2009) Tissue-specific cadmium accumulation and its effects on nitrogen metabolism in tobacco (*Nicotiana tabaccum, Bureley v.* Fb9). CR Biol 332:58–68

Electrochemical Synthesis of Multidimensional Nanoparticles and Their Photocatalytic Applications

Kalawati Saini, Smriti Sharma Bhatia and Nidhi Gupta

Abstract Copper and copper oxide nanoparticles (NPs) have been synthesized by electrochemical route using the Tri-Sodium Citrate 150 mM (TSC) as a capping and reducing agent. The synthesis has been done at 3.2 V, 311 K using copper rod as a working electrode and Platinum wire as a reference electrode. The electrochemical set up has been kept in air as well as under inert Nitrogen-purged conditions. The NPs have been characterized by using UV–visible absorption spectroscopy, Scanning electron microscopy (SEM), and powder X-ray diffraction (PXRD) techniques. This new kind of synthesis method shows the excellent stability compared with that of another chemical method of copper nanoparticles. These particles have been used in photocatalytic-oxidation of organic pollutants.

Keywords Tri-sodium citrate (TSC) · Working electrode · Platinum wire Reference electrode

1 Introduction

Copper and their oxide nanoparticles (NPs) especially cuprous oxide have wide application range as a catalyst such as in photo-electrochemical water splitting, photovoltaics, opto-electronics, biosensors, etc. The Cu_2O NPs have been used as a photo-catalyst for enhanced organic contaminant degradation under visible light irradiation [1]. The pH-Dependent single-step rapid synthesis of CuO and Cu_2O Nanoparticles have been done by using microwave irradiation [2]. These CuO nanoparticles displayed good hole mobility under ambient conditions making it a

K. Saini (✉) · S. S. Bhatia
Miranda House, University of Delhi, New Delhi 110007, Delhi, India
e-mail: kalawati.saini@mirandahouse.ac.in

N. Gupta
Kanoria PG Mahila Mahavidyalaya, University of Rajasthan, Jaipur 302004, India
e-mail: guptanidhi1@yahoo.com

V. S. Parmar et al. (eds.), *Green Chemistry in Environmental Sustainability and Chemical Education*, https://doi.org/10.1007/978-981-10-8390-7_5

good candidate to be used in FET devices. The NPs of CuO with flower like structure have been prepared by a domestic hydrothermal microwave [3]. Magnetic nanoporous Cu/(Fe,Cu)$_3$O$_4$ composites with excellent electrical conductivity have been investigated by one-step dealloying [4]. Cu and Cu-Based Nanoparticles with applications in Catalysis have been reviewed by Manoj B. Gawande. Copper/Copper-Oxide Nanoparticles has been synthesized and have been studied for optical properties [5]. The copper ion of antibacterial effect on *Pseudomonas aeruginosa*, *Salmonella typhimurium*, *Helicobacter pylori* and also reported for optical and mechanical properties respectively [6]. The graphene oxide/metal (Cu, Ni, Co) nanoparticle hybrid composites have been synthesized via a facile thermal reduction method [7]. Preparation of nanocrystalline Cu_2O thin film has been done by pulsed laserdeposition and sol-gel like dip technique [8, 9].

2 Results

UV–visible absorption spectra of copper nanoparticles in aqueous solution have been recorded which has been shown in Fig. 1. The characteristic absorption band at around 235–300 nm and 730 nm (surface-plasmon) have been found due to of Cu_2O and CuO nanoparticles respectively. The value of absorption band have been found same as reported in the literature. The sharpness of the absorption band for Cu_2O shows narrow size distribution of nanoparticles and the broaderness of the absorption band for CuO shows wide size distribution of nanoparticles. But the value for surface-plasmon of Cu nanoparticles has been not obtained in the range of 519–573 nm. Because we have taken the solution of electrolysis as it is, after applying potential of 3.2 V for two hours. The Cu nanoparticles are mainly deposited on platinum electrode. It means during electrolysis the Cu_2O nanoparticles remains in solution of electrochemical cell as well as deposited on platinum electrode. The PXRD pattern shows only formation of solid form of Cu nanoparticles and Cu_2O nanoparticles which is shown in Fig. 2a. It means overall all three

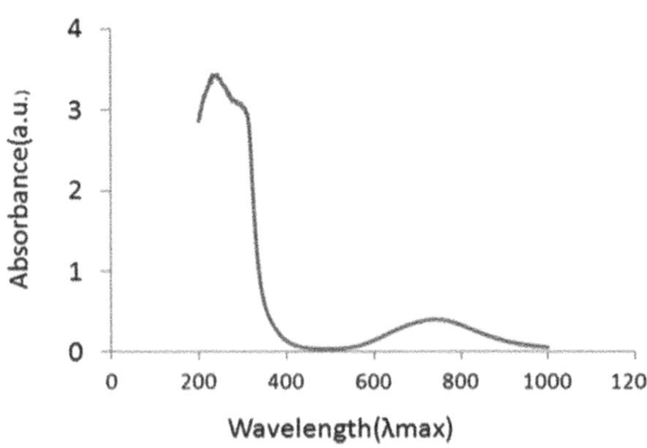

Fig. 1 UV-visible spectra of copper/copper oxide NPs prepared with 150 mM of tri-sodium citrate at 3.2 V, and 311 K

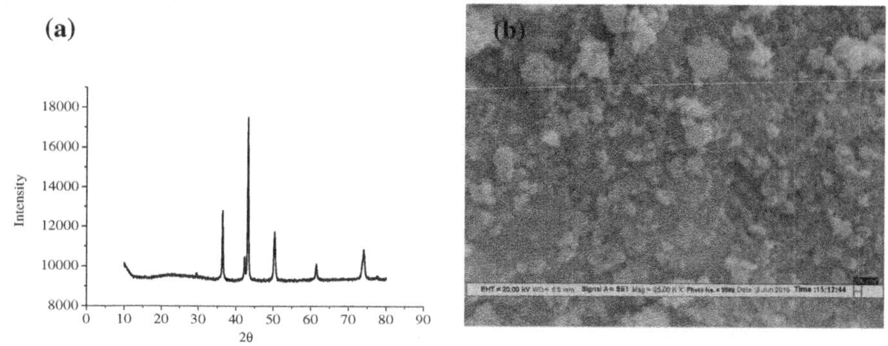

Fig. 2 **a** Powder XRD pattern of copper/copper oxide NPs prepared with 150 mM of tri-sodium citrate at 3.2 V, and 311 K and **b** SEM images of copper/copper oxide NPs prepared with 150 mM of tri-sodium citrate at 3.2 V and 311 K, Scale bar 200 nm

type of nanoparticles (Cu, Cu_2O, CuO) are synthesized via electrochemical route. But after electrolysis in solution phase both copper oxides are presented and on platinum electrode Cu and Cu_2O nanoparticles are deposited.

The XRD pattern has been recorded for the reddish brown powder deposited on reference electrode (Pt) after drying. The diffraction pattern shows peaks at 2θ values and corresponding (hkl) planes of 43.3° (111), 50.5° (200), 74.1° (220) that matches with the *fcc* metallic Cu (JCPDS, File No. 85-1326) and the other peaks in the same diffraction pattern at 2θ values of 29.58° (110), 36.44° (111), 42.3° (200), 61.4° (220), and 73.9° (311) that matches with the Cu_2O (JCPDS, File No. 65-328), indicating that Cu_2O also coexists together with metallic copper particles. These (hkl) values also match with the reported values in the literature [2]. The coexisting Cu_2O is also considered to be due to some oxidation in air environment in addition to Cu-hydroxide reduction process. No impurity diffraction peaks have been detected confirming the high purity of the product obtained by this method. Additionally, high reflection peak intensities for all the samples suggest that these are highly crystalline in nature.

Figure 2b shows the SEM images of copper/copper oxide nanoparticles prepared with 150 mM of TSC and an applied voltage of 3.2 V. This image shows that synthesized NPs are agglomerated in nature. The approximate size of particles has been found in the range of 50–60 nm. These particles are elongated and spherical in shape. These synthesized Cu/Cu_2O nanoparticles have been tested for photocatalytic activity.

2.1 Photocatlytic Degradation of Methylene Blue

Methylene Blue (MB) has been used for showing the photocatalytic activity of synthesized nanoparticles under UV light irradiations. The degradation of

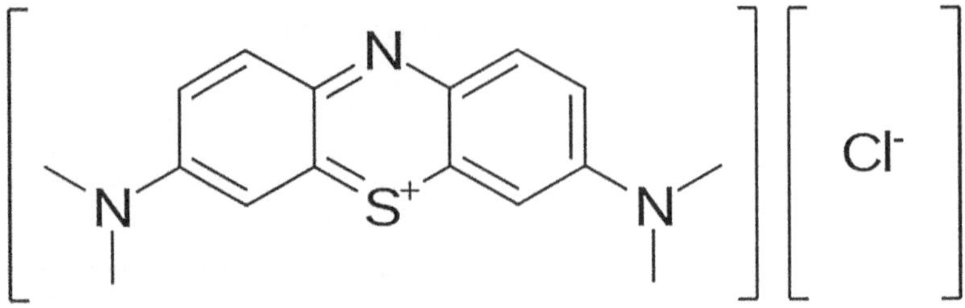

Scheme 1 Structure of Methylene Blue (MB) dye

methylene Blue in the presence of Cu NPs has been studied spectrotometrically by using Perkin-Elmer Lambda Bio 20 UV–vis spectrophotometer at a resolution of 1 nm.

The structure and mechanism of methylene blue have been given in Schemes 1 and 2.

This mechanism explains the nature of the intermediates leading to the low molecular weight hydrocarbon products. Eventually complete disintegration is expected with time to form the CO_2 and H_2O.

Measurements of the absorbance have been done at intervals of 3 minutes (min) over a time span of 30 min shows a decrease in the absorbance maximum of MB at 665 nm. For showing first order kinetics the curves ln C_t/C_0 versus time (min) have been drawn which is given in Fig. 3. Herein curve (a) shows degradation of MB with 150 mg of synthesized nanocatalyst, curve (b) with 125 mg and curve (c) with 100 mg of same synthesized catalyst in basic medium (0.1 M NaOH).

However, the rate of maximum reduction in the absorbance of methylene Blue has been obtained more prominent for the amount of nanoparticles of (150 mg) then the Copper and Copper oxide same nanoparticles of (125 mg) and (100 mg) respectively. The result show rapid increase in MB degradation from 0 to 30 min which becomes slower after 30 min for all three samples. The much higher MB degradation rate for the amount of 150 mg of catalyst proves its better performance compared to 125 mg and 100 mg of catalyst.

2.1.1 Mechanisms of Photodegradation

The following interactions represent the most common decomposition process adopted during the photodegradation of the organic dye by the synthesized Cu_2O nanocatalyst, like as given for TiO_2 catalyst in the literatrue.

$$Cu_2O + \text{dye} \rightarrow Cu_2O\text{---dye} \tag{1}$$

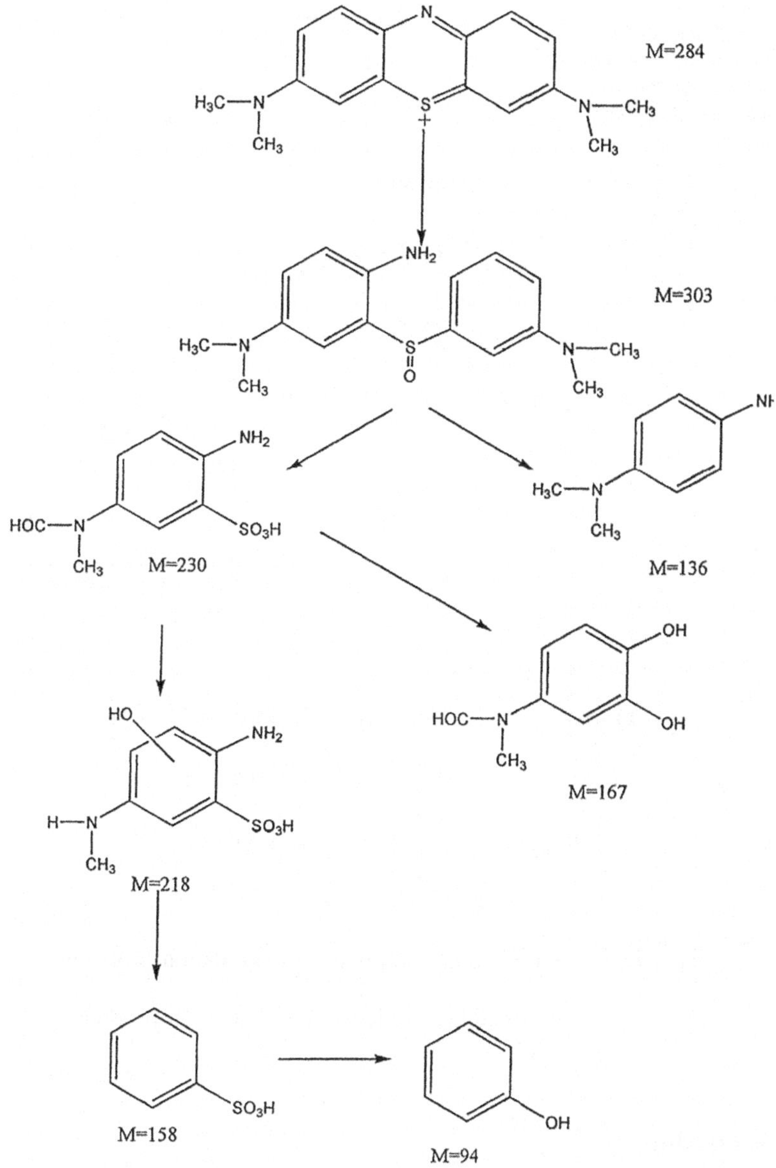

Scheme 2 Mechanism of photocatalytic degradation of MB over a photocatalyst under UV light irradiation

Fig. 3 Kinetics of photocatalytic degradation of methylene blue (in UV lamp photoreactor), experimental conditions: methylene blue (C_0) = 15 μM, Cu_2O = 150 mg (curve a), Cu_2O = 125 mg (curve b), Cu_2O = 100 mg (curve c)

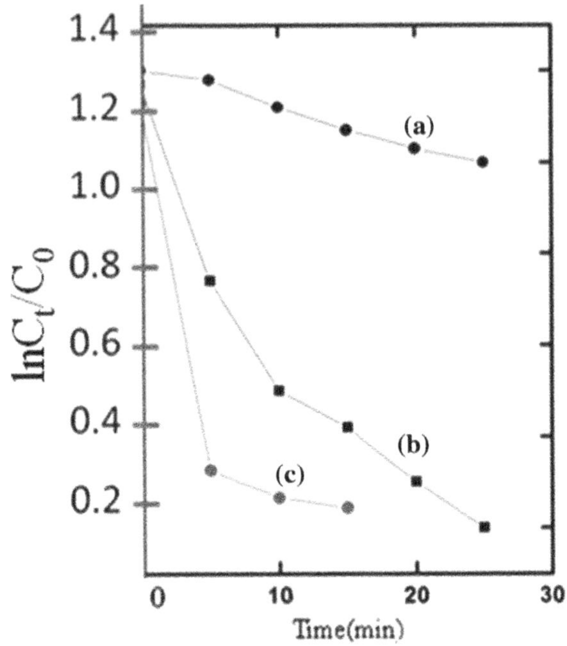

$$Cu_2O\text{--dye} + h\nu(\lambda < 390\ \text{nm}) \rightarrow Cu_2O(e^-, h^+)\text{--dye} \tag{2}$$

$$Cu_2O\,(h^+) + H_2O \rightarrow Cu_2O\,(OH^{\cdot}) + H^+ \tag{3}$$

$$Cu_2O\,(h^+) + OH^- \rightarrow Cu_2O\,(OH^{\cdot}) \tag{4}$$

$$Cu_2O(e^-) + O_2 \rightarrow Cu_2O(O_2^{-\cdot}) \tag{5}$$

$$OH^{\cdot} + \text{dye} \rightarrow \text{intermediate products} \rightarrow \text{degradation products} \tag{6}$$

$$O_2^{-\cdot} + \text{dye} \rightarrow \text{intermediate products} \rightarrow \text{degradation products.} \tag{7}$$

3 Discussion

The maximum absorbance has been determined at 665 nm for pure methylene Blue in basic medium. The degradation of methylene Blue has been compared with copper nanoparticles synthesized at 150 mM of tri-sodium citrate. The experiments have been performed for one cycle for all three samples of 15 μM methylene Blue with different amount of same catalyst. The colour of the reaction mixtures faded, indicating that degradation has been occurred.

Measurements of the absorbance have been done at intervals of 3 minutes (min) over a time span of 30 min shows a decrease in the absorbance maximum of MB at 665 nm (λ_{max}).

As we know that from literature that about 17% MB decomposes through photolysis under visible light irradiation with commercial TiO_2. It means it behaves weak catalyst in presence of visible light. As we know that in neutral or weakly acidic conditions, the dye degradation occurs mostly by the photocatalytic mechanism initiated by the photosensitization or photolysis. But in alkaline conditions, higher adsorption of MB on the catalyst surface occurs and due to this more decomposition of methylene Blue takes place.

The organic dyes in general are composed of carbon and heteroatom such as sulfur, nitrogen, chlorine etc. These elements are converted to a gas CO_2, and several ions such as nitrate ions, sulfate ions, and the ammonium ions as the photodegradation of the dye proceeds. In the case of the photodegradation of MB sulfate ions are formed through the formation of a sulfoxide, a sulfone and a sulfonic group as shown below.

$$OH^- + p^+ \rightarrow {}^{\cdot}OH \tag{8}$$

$$R - SO_3{}^- + {}^{\cdot}OH \rightarrow R - OH + SO_3{}^- \tag{9}$$

$$SO_3{}^- + {}^{\cdot}OH^- \rightarrow SO_4{}^{2-} + H^{\cdot} \tag{10}$$

The rate constant has been found to be 0.0314, 0.0505 and 0.1123 min^{-1} for the corresponding catalytic degradation of MB with 100 mg of Cu NPs, 125 mg of Cu NPs and 150 mg Cu NPs respectively. These results shows that the rate constant in presence of 150 mg of Cu NPs has been found more than 2.0 times than the 100 mg and 125 mg of Cu NPs for degradation of MB.

The results have been summarized in Table 1. It is evident from the curve given Fig. 3 that rate of degradation of MB has been obtained comparable with reported value in the literature for composite of TiO_2/C [10]. In our results the time taken for 87% degradation of MB is very less (30 min) with 150 mg of Cu NPs. Differences in catalytic performance with different amount of Cu/Cu_2O NPs can also be explained on the basis of the difference in the nanostructure-support contact area that depends on the particles shape and size. It is known that many catalytic processes occur at the perimeter interface around the nanoparticles where the fraction of step sites increases significantly with decreasing particle size.

Table 1 Comparative study of degradation of methylene Blue with three different amount of same copper/copper oxides NPs when the reaction is monitored for 30 min (Kinetics of pseudo first order)

S. no	Cu/Cu_2O with 150 mM TSC	Rate constant (min^{-1})
1	100	0.0314
2	125	0.0505
3	150	0.1123

4 Materials and Methods

Tri-Sodium Citrate capped copper nanostructures of (Cu, Cu_2O) have been synthesized via electrochemical route. Electrochemical cell has been formed by immersing the copper electrode and Platinum wire in solutions of 150 mM of Tri-Sodium Citrate (TSC). The solution has been prepared in double deionised water. The electrochemical set up has been kept in air as well as under inert Nitrogen-purged conditions equipped with a magnetic stirrer at 350 rpm. The powder present in the solution has been obtained by filtering with what -men paper of 42 number and has been washed several times by de-ionized water then dried in atmospheric condition and the reddishbrown powders deposited on platinum electrode can be easily separated. In this synthesis first copper gets oxidise in +2 state from working copper electrode when potential is applied or we can say dissolution process take place first after that Cu and Cu_2O form due to reduction process and size of particles has been controlled by capping agent.

4.1 Characterization

Nanoparticles have been characterized by using UV–visible absorption spectroscopy, Scanning electron microscopy (SEM) and X-ray diffraction(XRD) techniques. The UV-Visible absorption measurements have been done with Parkin Elmer UV/vis spectrophotometer LambdoBio20 at a resolution of 1 nm. X-ray diffraction measurements of the electroreduced dried powder has been done by filling it into the groove of quartz glass sample holder using Bruker made X-ray diffractometer instrument operating at a voltage of 40 kV and a current of 30 MA with Cu Kα radiation. The field emission scanning electron microscope (FE SEM) attached with Oxford-EDS system IE 250X max 80 with latest 80 mm^2 model FEI Quanta 200 have been used to determine morphology and elemental composition of the prepared materials. The photocatlytic activity has been studied by FLEXI UV–Visible Photo Chemical Triple Jacketed Reactor.

5 Conclusion

We have prepared Copper and copper oxide nanoparticles in different shape and size by a novel electrochemical synthesis technique which is simple and environment friendly. It is an easy, fast, and cost effective technique and doesn't involve any harmful and environmentally toxic chemicals used previously in conventional chemical reduction methods. The photoinduced hole oxidizes the methylene blue molecule to methylene blue radicals ($^{\cdot}$MB), which decompose fully to H_2O and carbon dioxide by further reaction with O_2. The photodegradation of methylene blue (MB) becomes more faster in presence of our synthesized copper/copper oxide than in presence of commercial TiO_2. The rate of decomposition increases when we increase amount of nanoparticles. These nanoparticales can also be used in photocatalytic degradation of toxic other organic pollutants such as Chlorophenol and methyl orange.

Acknowledgements The authors wish to thank to Department of Chemistry, Department of Textile, Indian Institute of Technology, Delhi, for providing facility for PXRD and SEM. The author wish to thank to Dr. Pravin P. Ingole, Assistant Professor, Department of Chemistry, Indian Institute of Technology Delhi for providing guidance during this research work.

Conflicts of Interest Declare conflicts of interest or state "The authors declare no conflict of interest."

References

1. Liu I, Liang Y, Cui W (2016) Cu_2O NPs decorated $BiPO_4$ photo-catalyst for enhanced organic contaminant degradation under visible light irradiation. RSC Adv 6:29202–29209
2. Nikam AV, Arulkashmir A, Krishnamoorthy K, Kulkarni AA, Prasad BLV (2014) pH-dependent single-step rapid synthesis of CuO and Cu_2O nanoparticles from the same precursor. Cryst Growth Des 14:4329–4334
3. Volanti DP, Keyson D, Cavalcante LS, Simoes AZ, Joya MR, Longoa E, Varela JA, Pizani PS, Souza AG (2008) Synthesis and characterization of CuO flower-nanostructure processing by a domestic hydrothermal microwave. J Alloys Compd 459:537–542
4. Gawande MB, Goswami A, Felpin FX, Asefa T, Huang X, Silva R, Zou X, Zboril R, Varma RS (2016) Cu and Cu-based nanoparticles: synthesis and applications in catalysis. Chem Rev 116:3722–3811
5. Kim YH, Choi YR, Kim KM, Choi SY (2012) Evaluation of copper ion of antibacterial effect on *Pseudomonas aeruginosa*, *Salmonella typhimurium* and *Helicobacter pylori* and optical, mechanical properties. Appl Surf Sci 258:3823–3828
6. Huang X, Zhao G, Wang X (2015) Fabrication of reduced graphene oxide/metal (Cu, Ni, Co) nanoparticle hybrid composites via a facile thermal reduction method. RSC Adv 5:49973–49978
7. Qi Z, Gong Y, Zhang C, Xu J, Wang X, Zhao C, Ji H, Zhang Z (2011) Fabrication and characterization of magnetic nanoporous Cu/(Fe, Cu)3O4 composites with excellent electrical conductivity by one-step dealloying. J Mater Chem 21:9716–9724
8. Jawad MF, Ismail RA, Yahea KZ (2011) Preparation of nanocrystalline Cu_2O thin film by pulsed laser deposition. J Mater Sci 22:1244–1247
9. Ray SC (2001) Preparation of copper oxide thin film by the sol–gel-like dip technique and study of their structural and optical properties. Sol Energ Mat Sol Cells 68:307–312
10. Ramli ZAC, Asim N, Isahak, WNRW, Emdadi Z, Ludin NA, Yarmo MA, Sopian K (2014) Photocatalytic degradation of methylene blue under UV light irradiation on prepared carbonaceous TiO2. Sci World J 415136:1–8. http://dx.doi.org/10.1155/2014/415136

Pine Needles as Green Material for Removal of Metal Ions and Dyes from Waste Water

Kanika Gulati, Christine Jeyaseelan and Sunita Rattan

Abstract Carcinogenic properties of Chromium(VI) have been studied extensively and reported. The majority of the environmental releases of chromium are from industrial sources. Rhodamine B is a synthetic dye widely used as colourant in the manufacture of textiles and food stuffs. Rhodamine dyes present in drinking water could lead to subcutaneous tissue borne sarcoma which is highly carcinogenic. Other toxicities like reproductive and neurotoxicity have been investigated and proved. The work reported involves the removal of Cr(VI) as well as Rhodamine B dye using pine needles by the process of adsorption. The adsorption studies were performed by batch process and the detection done using UV–visible spectrophotometer at 540 nm for Cr(VI) and 554 nm for Rhodamine B. Various parameters like pH, concentration of adsorbent, mass of adsorbate, time of contact were optimized. The percentage efficiency for removal under optimized conditions, of Cr (VI) and Rhodamine B dye was up to 96 and 92% respectively. The adsorption studies were performed using Freundlich and Langmuir adsorption isotherms as well as the kinetic and thermodynamic parameters have also been determined. This technique can be effectively used for the detoxification of water by the removal of metal ions and dyes obtained from industries.

Keywords Spectrophotometry · Hexavalent chromium · Rhodamine B
Pine needles

1 Introduction

Compounds of Chromium are commonly used for mining, metallurgical and dyeing industries [1] from where it enters the soil and water. The ions of Chromium(III) is not very toxic and stable but Chromium(VI) is toxic. Cr(VI) in the form of CrO_4^{2-} or $HCrO_4^-$ diffuses through cell membranes. The high toxicity of Chromium(VI) is

K. Gulati · C. Jeyaseelan (✉) · S. Rattan
Amity Institute of Applied Sciences, Amity University, Noida, India
e-mail: cjeyaseelan@amity.edu

V. S. Parmar et al. (eds.), *Green Chemistry in Environmental Sustainability and Chemical Education*, https://doi.org/10.1007/978-981-10-8390-7_6

of great concern as it affects human health significantly. It enters the blood and may cause damage to the liver, kidneys etc. due to its action as an oxidizing agent [2]. The carcinogenic nature of Cr(VI) is well known [3]. According to the World health organization, 0.05 mg/l is the maximum allowable concentration [4].

The textile industry is one of the biggest users of dyes. Rhodamine B is a synthetic dye used as a colourant for the manufacture of textiles and food stuff. This dye has been proved to be highly carcinogenic. Reproductive effects and neuro-toxicity have been investigated and proved, due to exposure to this dye [5].

Literature reports various techniques for the removal of dyes and metal ions like electrochemical reduction, chemical precipitation, electrodialysis, ultrafiltration, ion exchange etc. [6]. These methods have their own disadvantages like high cost especially for large scale water treatment. The more effective technique for their removal has been using the process of biosorption. This method involves the removal of the pollutants by using biodegradable materials like wheat residue [7], agricultural biowastes [8], tea waste [9], chitosan [10] etc. Synthetic materials do not have a good biocompatibility and biodegradability. However, the natural materials also have limitations like process ability, stability etc. The presence of groups like hydroxyl, carbonyl, sulphahydryl, phosphoryl etc. in plant biomass have the ability to adsorb metal ions and dyes [11, 12].

The present work aims at using naturally abundant pine needles which is a waste material available easily in the mountainous regions for effectively removing some pollutants-metal ions like Cr(VI) and dyes like Rhodamine B. The various parameters have been optimized and the adsorption isotherms and the kinetic studies have also been reported.

2 Results

2.1 *Effect of pH*

A known concentration of Cr(VI)/Rhodamine B was taken in a conical flask along with a given amount of adsorbent and the pH was set from 1 to 10 using 0.5 M HNO_3 and 0.1 N NaOH. The percentage removal is shown in Fig. 1. The maximum recovery of Chromium(VI) was observed at pH 1 and for Rhodamine B at pH 4.

2.2 *Effect of Adsorbent Dosage*

The amount of adsorbent added was varied and its effect on the removal of Cr(VI) and Rhodamine B was studied. The observation was plotted as in Fig. 2. It was observed, for the recovery of Cr(VI) an amount of 0.4 g of adsorbent was optimum

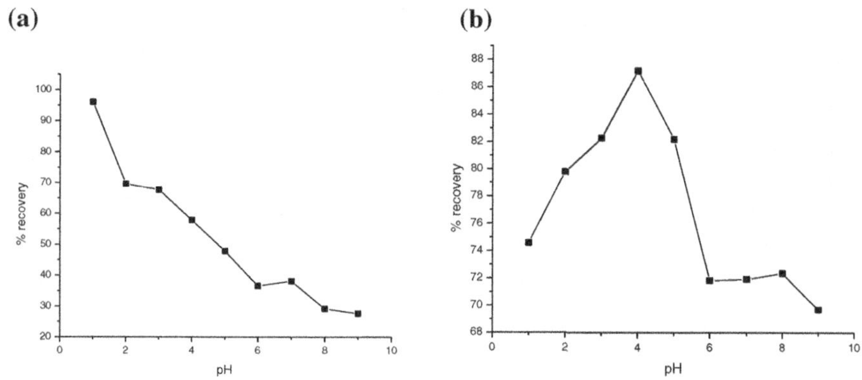

Fig. 1 Effect of pH **a** Chromium(VI), **b** Rhodamine B

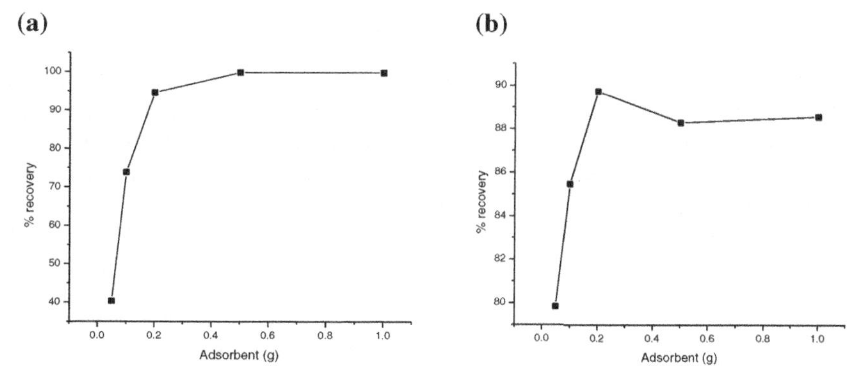

Fig. 2 Effect of adsorbent dosage **a** Chromium(VI), **b** Rhodamine B

and for Rhodamine B 0.2 g of the adsorbent gave the maximum recovery. There was no observable increase in the percentage removal at higher amounts of adsorbent due to saturation in the adsorption sites.

2.3 *Effect of Contact Time*

The effect on adsorption with the time of contact of adsorbent with adsorbate was studied from 30 to 200 min which is represented in Fig. 3. The study showed that with increase in contact time the amount of Cr(VI) removed was maximum at 180 min after which it became constant. In case of Rhodamine B the percentage removal was not affected by time and was constant after 30 min.

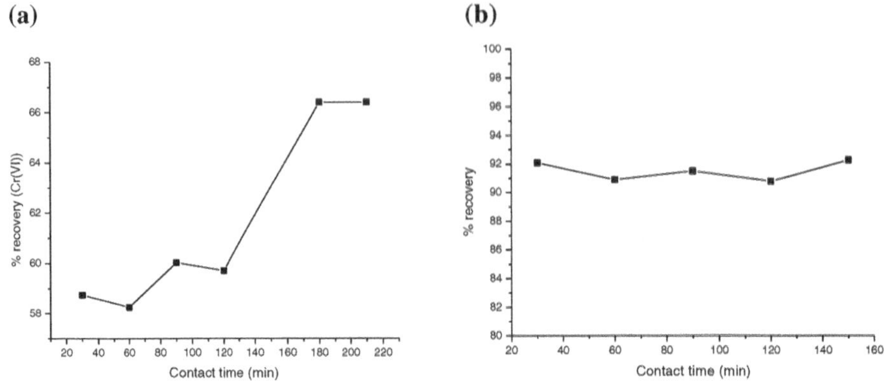

Fig. 3 Effect of contact time **a** Chromium(VI), **b** Rhodamine B

2.4 *Effect of Adsorbate Concentration*

Solution with fixed concentration of Cr(VI) was prepared from 2 to 12 mg/l keeping the adsorbent dosage fixed. Similarly the effect of changing the concentration of Rhodamine B from 0.1 to 1.0 mg/l was analyzed. The results obtained are shown in Fig. 4a, b respectively. The data obtained showed that the percentage recovery decreased with increasing concentration.

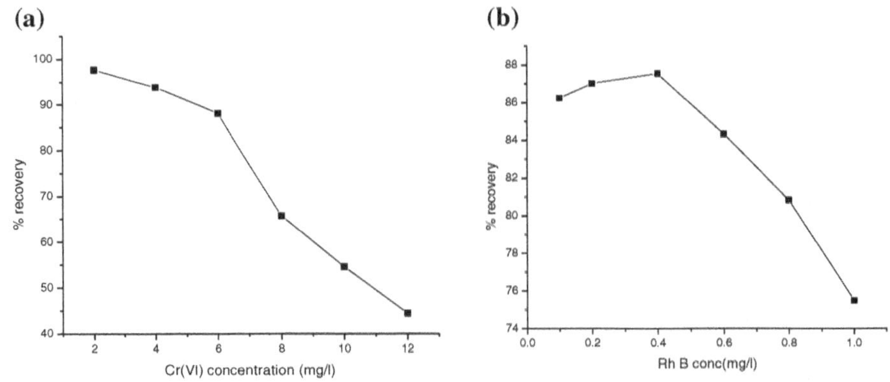

Fig. 4 Effect of concentration **a** Cr(VI), **b** Rhodamine B

2.5 *Effect of Temperature*

The increase in temperature had an effect on the adsorption of Cr(VI) and Rhodamine B which was studied at 298, 303 and 313 K. In both the cases, there was a decrease in the percentage removal with increase in temperature.

2.6 *Adsorption Isotherms*

The adsorption Isotherms were studied quantitatively to know the kind of adsorption process taking place.

2.6.1 Langmuir Adsorption Isotherm

This model showed monolayer adsorption taking place on a homogenous surface [13]. It shows a relationship between the maximum adsorption capacity (q_0) and adsorption energy (b) as given below:

$$C_e/q_e = 1/q_0 b + C_e/q_0$$

A graph was plotted between C_e/q_0 against C_e. The intercept and slope gave value of q_0 and constant b respectively.

The Langmuir adsorption Isotherms are shown in Fig. 5.

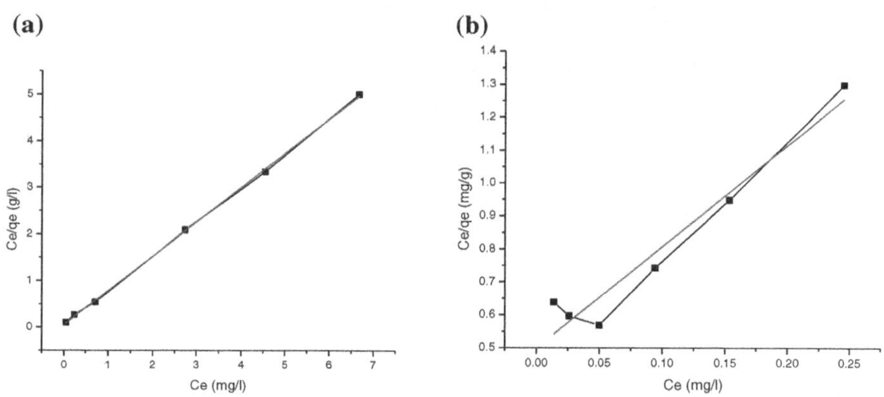

Fig. 5 Langmuir adsorption isotherm **a** Cr(VI), **b** Rhodamine B

2.6.2 Freundlich Adsorption Isotherm

This isotherm explains the adsorption on inequivalent adsorption sites due to the surface being heterogenous [14]. The isotherm equation is given as:

$$\log q_e = \log K_F + 1/n \log C_e$$

where K_F is adsorption capacity and n indicates the intensity of adsorption. The graph between log q_e versus log C_e was used to calculate the value of K_F and n from the slope and intercept.

The Freundlich adsorption isotherms are shown in Fig. 6.

2.7 *Adsorption Kinetics*

The kinetics of adsorption is relevant to this study because the amount of adsorbate adsorbed by the adsorbent is dependent on the time of contact. The pseudo first order and pseudo second order kinetic models were followed to study solid-liquid adsorption.

The pseudo first order kinetics [15] is as given below

$$\log(q_e - q_t) = \log(q_e) - k_1 t/2.303$$

where q_e (mg/g) is the amount of Cr(VI)/dye adsorbed at equilibrium and q_t (mg/g) is the amount of Cr(VI)/dye adsorbed at at time t with a rate constant k_1 (min^{-1}). The graphs obtained are as shown in Fig. 7.

The pseudo second order equation [16] is given as

$$t/q_t = 1/k_2 q_e^2 + t/q_e$$

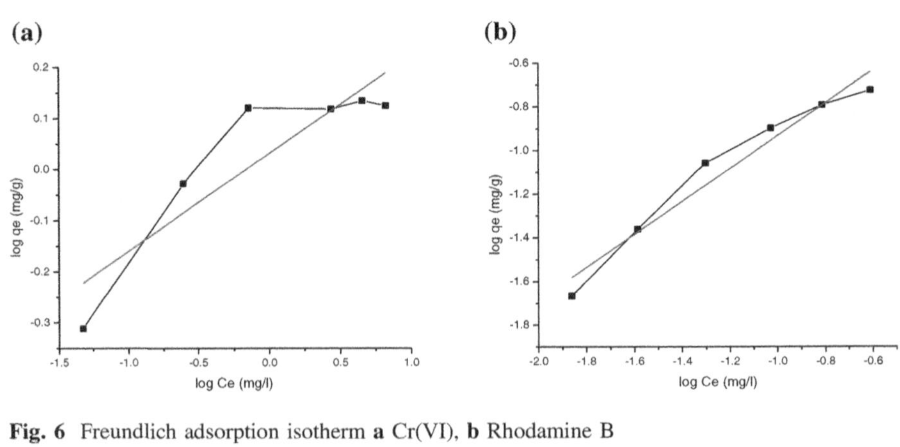

Fig. 6 Freundlich adsorption isotherm **a** Cr(VI), **b** Rhodamine B

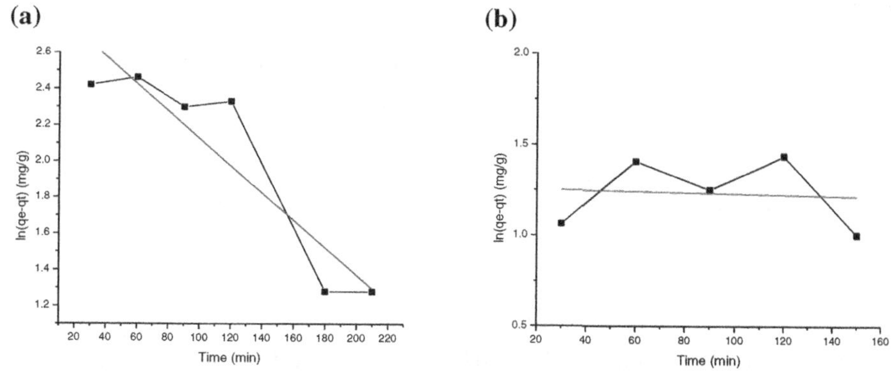

Fig. 7 Pseudo first order kinetics **a** Chromium(VI), **b** Rhodamine B

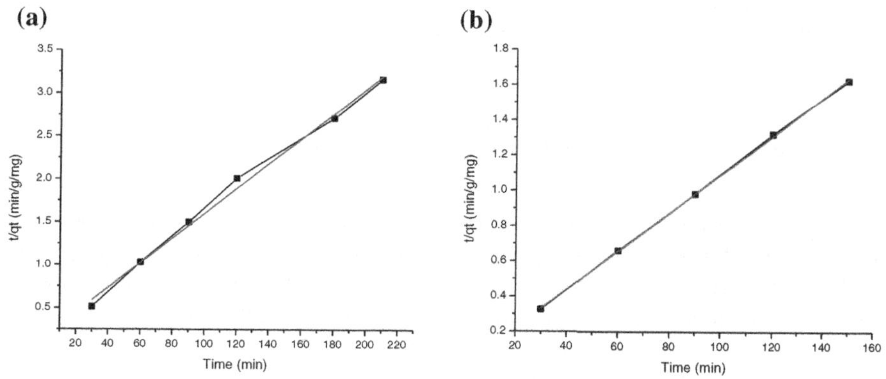

Fig. 8 Pseudo second order kinetics **a** Chromium(VI), **b** Rhodamine B

where k_2 is obtained from the plot of log t/q_t versus t which is the pseudo second order rate constant. The graphs obtained are as shown in Fig. 8.

2.8 Thermodynamics of Adsorption

The free energy change (ΔG°) is determined as

$$\Delta G^\circ = -RT \ln K$$

Table 1 Thermodynamic parameters

S. No	Adsorbate	Temperature (K)	ΔG° (kJ/mol)	ΔH° (kJ/mol)	ΔS° (kJ/K/mol)
1	Cr(VI)	298	−2900.63	−53.79	−0.17
		303	−1986.24		
		313	−325.71		
2	Rhodamine B	298	−339.17	−7.35	−0.025
		303	−492.53		
		313	−731.76		

The entropy (ΔS°) and enthalpy (ΔH°) are calculated by van't Hoff equation as

$$\ln K = \Delta S^\circ/R - \Delta H^\circ/RT$$

(R is gas constant (8.314 J/K/mol).

The data obtained are as shown in Table 1.

2.9 *Regeneration and Reusability of Adsorbent*

From a green chemistry view reagents like sodium hydroxide, hydrochloric acid and sodium chloride were used for desorption studies. It was observed that NaCl gave the best results. The regenerated adsorbent could be used with the same adsorption efficiency for 3 cycles with slight decrease in efficiency.

3 Discussion

The study showed that the removal of Chromium(VI) and Rhodamine B dye from the solution is dependent on various factors. From the effect of pH it was observed that at low pH, chromium exists in the form of $HCrO_4^-$ while at high pH value, it is found in various other forms like $Cr_2O_7^{2-}$, $Cr_3O_{10}^{2-}$, $HCrO_4^-$. But at pH 1, $HCrO_4^-$ is the dominant species. As the pH is increased, the equilibrium shifts from $HCrO_4^-$ to $Cr_2O_7^{2-}$ and CrO_4^{2-} [17]. The adsorbent surface may be surrounded by hydronium ions. This may lead to an increase the interaction of Cr(VI) due to the increase in the binding strength of the sites on adsorbent. With the increase in pH, the total surface gets a negative charge and thus adsorption decreases [18]. The equations for the equilibrium in aqueous solution of Chromium(VI) anion can be stated as: [19].

$$H_2CrO_4 \leftrightarrow HCrO_4^- + H^+ \quad k_1 = 1.21$$

$$Cr_2O_7^{2-} + H_2O \leftrightarrow 2HCrO_4^- \quad k_2 = 35.5$$

$$HCrO_4^- \leftrightarrow CrO_4^{2-} + H^+ \quad k_3 = 3 \times 10^{-7}$$

At pH values of greater than 6 the adsorption was not significant due to complexation of CrO_4^-, $Cr_2O_7^{2-}$ and OH^- ions which gets adsorbed at the surface and the adsorption of OH^- is predominant.

The effect of adsorbent dosage showed that once all the sites are saturated, the percentage removal becomes constant. The effect of contact time showed that the adsorption is a two stage process. In the initial stage the adsorption takes place at a faster rate while in the next stage it is a slower process. This maybe because in the starting phase, there are a large number of vacant sites, which led to faster adsorption, but once the vacant sites are exhausted the process slows down. There is a probability of repulsion between the solute molecules in aqueous phase and bulk phase [20]. On varying the concentration of the adsorbate it was seen that at lower concentration, a maximum adsorption is obtained due to maximum interaction with the binding sites. But at higher concentration the percentage removal decreases due to the shortage of binding sites.

The isotherm studies showed that the linearity was much more in case of Langmuir than Freundlich in both cases (since the R^2 value was closer to 1). The values obtained are as given in Table 2. Higher value of n indicates active interaction between adsorbent and adsorbate.

From the kinetic study it was observed that it followed pseudo second order (since the R^2 value was close to 1) (Table 3).

The value of $\Delta G°$ comes out to be negative which shows that the process is spontaneous, the exothermic nature is shown by the negative enthalpy change while the decrease in randomness is shown by the negative entropy change the adsorbent goes from the solution to the adsorbed state.

Table 2 Linear regression data for Langmuir and Freundlich isotherms

S No	Adsorbate	Langmuir isotherm			Freundlich isotherm		
		R^2	q_m (mg/g)	K_L (L/mg)	R^2	n	K_F
1	Cr(VI)	0.999	1.357	14.92	0.761	5.2	1.07
2	Rhodamine B	0.957	0.324	6.188	0.942	1.33	0.794

Table 3 Linear regression data for pseudo first order and pseudo second order kinetics

S No	Adsorbate	Pseudo first order			Pseudo second order		
		R^2	K_{ad}	q_e	R^2	K_{ad}	q_e
1	Cr(VI)	0.807	0.001	741.3	0.999	3.003	17.7
2	Rhodamine B	−0.321	0.0013	69.44	0.999	0.01	100

4 Materials and Methods

4.1 Materials

Diphenyl carbazide(DPC), sodium hydroxide(NaOH), Rhodamine B dye, potassium dichromate and sulphuric acid (H_2SO_4) were procured from Merck. Double distilled water was used to prepare the aqueous solutions. Pine needles were collected from Himachal Pradesh district. The purity of the reagents utilized for the entire study were of analytical grade and did not need further purification.

4.2 Preparation of Adsorbent

The pine needles which had fallen from the trees were collected and the impurities like dust etc. was removed by washing a 4–5 times with distilled water. After this the needles were washed with hot water till a colourless filtrate was obtained. The needles were then dried in an over at 105 °C for about 2 h after which they were crushed and finely powdered which was then sieved to get a uniform particle size. The powder form was then used for further adsorption experiments without further modification. The needles were dried and crushed powder and stored in polythene bags to prevent absorption of moisture from the atmosphere.

4.3 Preparation of Adsorbate

2.82 g of potassium dichromate ($K_2Cr_2O_7$) was dissolved in 1 L of distilled water to prepare the stock solution and 0.1 g of Rhodamine B was dissolved in 100 ml distilled water. The Cr(VI) solution was then diluted 10 times and Rhodamine B 100 times for actual studies.

4.4 Batch Adsorption Studies

All the studies were carried out in a 250 ml flask with the required amount of adsorbent and the respective solution concentration. The flasks were then shaken by placing in a shaker at a fixed agitation speed to allow it to attain equilibrium. Diphenyl carbazide gave a violet coloured complex with Cr(VI) which showed an absorbance maxima at 540 nm. Rhodamine dye gave a maximum absorbance at 554 nm. The spectrophotometric studies were carried out on Schimadzu UV spectrophotometer-1800. Systronics digital pH meter-335 was used to measure the pH. The effect of various parameters like contact time, initial chromium concentration, adsorbent dosage, pH etc. was studied.

The percentage of removal of Chromium(VI) was determined by:

$$\%\,\text{Chromium removal} = \text{C}_\text{i} - \text{C}_\text{f}/\text{C}_\text{i} * 100$$

where C_i is the concentration before adsorption and C_f is the concentration after adsorption.

The adsorption Isotherms were plotted and kinetic studies were performed as well as thermodynamic parameters were calculated. All the experiments were repeated three times to get reliable results.

5 Conclusion

In this study it was observed that pine needles have very good adsorption capacity for hexavalent Chromium as well as Rhodamine B. Pine needles gave a maximum adsorption of 96% for Cr(VI) at pH 1 on addition of 0.4 g of adsorbent with contact time of 180 min while with Rhodamine B the recovery was 92% at pH 4 with contact time of 30 min on the addition of 0.2 g of adsorbent. Adsorption studies showed that Langmuir model fit better than Freundlich. The kinetic study showed that both Chromium(VI) and Rhodamine B followed pseudo second order kinetics. The spontaneity and exothermic nature of the process is shown by the thermodynamic parameters.

Acknowledgements The authors are grateful to Amity University for providing the necessary infrastructure and facilities for performing the experiments and doing the study.

Conflicts of Interest The authors declare no conflict of interest.

References

1. Gandhi MR, Vishwanathan N, Meenakshi S (2010) Preparation and applications of alumina/chitosan biocomposite. Int J Biol Macromol 47:146–154
2. Katz SA, Salem H (2006) The toxicology of chromium with respect to its chemical speciation: a review. J Appl Toxicol 13:217–224
3. Dayan AD, Paine AJ (2001) Mechanism of chromium toxicity, carcinogenicity and allegenicity; review of literature from 1985 to 2000. Hum Exp Toxicol 20:439–451
4. Basketter D, Horev L, Slodovnik D, Merimes S, Trattner A, Ingber A (2000) Investigation for the threshold for allergic reactivity to Chromium. Contact Dermat 44:70–74
5. Jain R, Mathur M, Sikarwar S, Mittla A (2007) Removal of hazardous dye Rhodamine B through photocatalytic and adsorption treatment. J Environ Manag 85:956–964

6. Owlad M, Aroua MK, Daud WAW et al (2009) Removal of hexavalent chromium-contaminated water and waste water: a review. Water Air Soil Pollut 200:59–77
7. Wang XS, Chen LF, Li FY, Chen KL, Wan WY, Tang YJ (2010) Removal of Cr(VI) with wheat residue derived black carbon: reaction mechanism and adsorption performance. J Hazard Mater 175:816–822
8. Park D, Lim SL, Yun YS, Park JM (2008) Development of new Cr(VI) biosorbent from agriculture biowaste. Bioresour Technol 99:8810–8818
9. Malkoc E; Nuhoglu (2006) Fixed bed studies for the sorption of chromium(VI) onto tea factory waste. Chem Eng Sci 61:4363–4372
10. Crini G (2005) Recent developments in polysaccharide based materials used as adsorbents in wastewater treatment. Prog Polym Sci 30:38–70
11. Demiral H, Demiral I, Tumsek F, Karabacakoglu B (2008) Adsorption of chromium(VI) from aquesous solution by activated carbon derived from olive bagasse and applicability of different adsorption models. Chem Eng J 144:188–196
12. Park D, Lim SR, Yun YS, Park JM (2007) Reliable evidences that the removal mechanism of hexavalent chromium by natural biomaterials is adsorption-coupled reduction. Chemosphere 70:298–305
13. Langmuir I (1918) The adsorption of gases on plain surface of glass, mica and platinum. J Am Chem Soc 40:1361–1403
14. Freundlich HMF (1906) Over the adsorption in solution. Z Phys Chem 57:385–470
15. Lagergren, S (1898) Zur theorie der sognnanten adsorption geloster Stoffe. K Sven Vetenskapsakad Handl 24:1–39
16. Ho YS (2006) Review of second order models for adsorption systems. J Hazard Mater 8136:681–689
17. Bansal M, Garg U, Singh D, Garg VK (2009) Removal of Cr(VI) from aqueous solutions using pre consumer processing agricultural waste. J Hazard Mater 162(1):312–320
18. Malkoc E, Nuhoglu Y, Abali Y (2006) Cr(VI) by waste acorn quercus ithaburensis in Fixed Beds: prediction of breakthrough curves. Chem Eng J 119(1):61–68
19. Miretzkya P, Cirelli AF (2010) Cr(III) and Cr(VI) removal from aqueous solution by raw and modified lignocellulosic material: a review. J Hazard Mater 180:1–19
20. Qaiser S, Saleemi AR, Umar M (2009) Biosorption of lead from aqueous solution by ficus religiosa leaves. Batch and column study. J Hazard Mater 166(2–3):998–1005

Pellet Biochar: An Environmental Remedy

Manoj Kumar, Amit Kumar Rawat, Anand Sonkar, Amar Kumar, Anusha Azhar, Mallika Kumar and Mohammad Mujahid Ali Khan

Abstract With recent impetus on building of smart cities, modern urban settlements and expansion of established settlements, solid waste management, sufficient energy supply, water pollution and water borne diseases have become an inseparable trouble. The need of the hour, thus, is an efficient and rapid remedy to these issues which is not only feasible but economically viable as well. Present study concentrates on how horticultural and commercial organic waste can be converted to pellets and after further processing can be utilised to combat water pollution and production of energy. Pellets are condensed product of above mentioned waste which is capable of utilizing these waste that is generated each day in tons. Pellets can be utilized directly as an alternate mode of fuel in certain areas. Analysis of pellets makes it clear that it's a cleaner and greener substitute to coal. Pellets after combustion gives rise to biochar which finds wide application in water purification and hence is considered as a remedy to water pollution. Pellet biochar possesses certain specific architecture and molecular composition which varies with temperature, residual time, feedstock type

M. Kumar · A. K. Rawat · A. Sonkar (✉) · A. Azhar · M. Kumar
Hans Raj College, University of Delhi, New Delhi, Delhi, India
e-mail: anandsonkar@gmail.com

M. Kumar
e-mail: kushwah205@gmail.com

A. K. Rawat
e-mail: amitrawat084@gmail.com

A. Azhar
e-mail: anushakhan1696@gmail.com

M. Kumar
e-mail: mallikakumar74@gmail.com

A. Kumar
Daulat Ram College, University of Delhi, New Delhi, Delhi, India
e-mail: amarleo123@gmail.com

M. Mujahid Ali Khan
Faculty of Engineering and Technology, University Polytechnic,
Aligarh Muslim University, Aligarh, India
e-mail: mujahidchemistry@gmail.com

V. S. Parmar et al. (eds.), *Green Chemistry in Environmental Sustainability and Chemical Education*, https://doi.org/10.1007/978-981-10-8390-7_7

and heat transfer rate. On gasification of pellets, syngas is produced which can be used for running engines as well as generating electricity.

Keywords Solid waste management · Energy · Pollution · Fuel Calorific value · Pellet biochar · Pyrolysis · Gasification

1 Introduction

Harbouring a population of 1027 million, as per 2001 census, India is a country of 5161 towns and cities that inhabited by 285 million people [1]. The total quantity of solid waste generated in urban area per day is *ca* 1.15 lakh tons which narrows down to 21,100 tonnes per day as in the case of six megacities in India [1]. Out of this, 30–35% [1] of the waste is compostable or biodegradable that is horticultural and organic waste which on prolonged, untreated exposure in the landfills emits methane due to the action of various microorganism on them. The emission of methane gas triggers fire in the landfills and raise the level of pollution that augments health hazards. This problem can be countered if the huge amounts of solid waste can be condensed to a smaller amount. There are a number of ways in which this can be done, one of them is pelletization.

Pellets are biofuels made from compressed organic matter or biomass. Pellets can be made from any one of five general categories of **biomass**: industrial waste and co-products, food waste, agricultural residues, energy crops, and virgin lumber. Present study deals with the pellet production primarily from the horticultural and organic waste. Pellets are in itself a versatile product and its conversion to biochar make them not only more useful but also a remedy for various environmental issues. The horticulture and organic waste is thus converted from a possible environmental hazard to an environmental remedy.

Biochar is a solid material obtained from the carbonisation of biomass. Biochar may be added to soils with the intention of improving soil function and to reduce emissions from biomass that would otherwise naturally degrade the environment by adding to greenhouse gases. Biochar also has appreciable carbon sequestration value. These properties are measurable and can be verified in a characterization scheme or in a carbon emission offset protocol [2]. The origin of biochar is associated to the Amerindian populations in the Amazon region, locally known as Terra Preta de Indio, where dark earth was created through the technique of *slah* and *char* techniques. Research on Terra Preta soils in Amazonia ash revealed the effects of biochar on fictionalization of soils. Especially, because the biochar has been known as an excellent soil amender for soil fertility and sustainability, many researchers and farmers worldwide are paying attention to its hidden secrets. Biochar is recognized as a very significant tool for environmental management [3]. It is a porous carbon adsorbent material with a potential to retain contaminants, also plays a crucial role in risk management and disposal of pollutants [4–8]. Lately, biochar was introduced as a bio-adsorbent material for removal of various contaminants in effluents, such as heavy metals, pigments, dyes, naphthalene, phosphorus, 1-naphthol, atrazine, and some macro- and micronutrients [9].

The present study outlines the methodology that needs to be followed for optimum pellet production, the versatile use of pellets, its conversion to biochar.

2 Results

Six types of pellets were made. The pellets that were composed of lawn grass and saw dust along with starch water as binder were identified to posses the strength and durability required for using it as a fuel directly and conversion to biochar. Moisture, fixed carbon content and macroelements were analysed. The chosen pellets have been referred to as grass pellets.

2.1 Physical and Chemical Properties of Grass Pellets

2.1.1 Energy Report of Pellets

The calorific value is the amount of energy stored in a sample which on combustion is released as thermal energy. The calorific value, therefore, dictates the desirability of a sample as a fuel. Currently, Cow dung cakes, coal and wood are the major cooking fuels used in certain parts of rural India. The calorific value of grass pellets (3788 kcal/kg) is higher than cow dung cake (3140 kcal/kg) and is comparable to those of wood (3500 kcal/kg) and Indian Coal (4000 kcal/kg) however it is much lower than that of the LPG (11,840 kcal/kg).

2.1.2 Moisture Content

Moisture content is an essential element in determining the energy of a fuel. High moisture content tends to lower the calorific value and thus, reduces the efficiency of the fuel. The optimum value of moisture in pellets is between 10 and 15%. However, the moisture of grass pellets can be influenced during pelletisation process and the desired content can be controlled during the inceptive steps. The content of moisture in the pellets produced during the experiment was found to be 11.4%.

2.1.3 Fixed Carbon Content

Fixed carbon is the solid residue left after the volatile materials have been expelled from the sample. Deduction of percentages of ash, volatile matter and moisture from the original mass of the sample yields the fixed carbon content. A sample with high fixed carbon content requires long combustion time. Grass pellets (90% *Cynodon* grass) have substantially lower value when compared with cow dung cakes, wood and Indian coal.

2.1.4 Macroelement Content in Pellets

The ash, obtained after combustion of the lignocellulosic pellets, contain higher level of Nitrogen, Phosphorus and Potassium than cow dung cakes, wood ash and Indian coal. This makes them a potential low-grade fertilizer.

2.2 *Tables*

See Tables 1 and 2.

Table 1 Comparison of physical properties of grass pellets with other conventional fuel

	Pellets	Cow dung	Wood	Indian coal
Calorific value (kcal/kg)	3788	3140	3513.2	4000
Moisture content (% by mass)	11.4	4.3	20	5.98

Table 2 Comparison of chemical composition of pellets with other conventional fuel

		Pellets	Cow dung	Wood	Indian coal
Amount of fixed carbon (%)		5.14	13.30	17–25	34.69
Macroelement content (% by mass)	Nitrogen	1.28	0.97	0.15	1.22
	Phosphorous	3.13	0.69	0.53	0.05
	Potassium	9.77	1.66	2.6	1.1

3 Discussion

Experiments and analysis carried out so far has shown that calorific value of grass pellets comes out to be 3788 kcal/kg which is quite comparable to the calorific values of wood (4600–4800 kcal/kg) and coal (4000–7000 kcal/kg). Grass pellets have higher calorific value than Indonesian coal (2384 kcal/kg). Fixed Carbon content for grass pellet is 5.14% which is 1/8th value of coal (40%) and half the value of wood (12%). Extra element detection while burning the grass pellet came out to be negligible. Sulphur in fuel, when combusted, produces Sulphur oxides (SOX) which are acid rain causing chemicals. Grass pellet has very low amount of Sulphur content in them as compared to wood and coal as shown by pellet analysis. Greenhouse gas emissions impacts our future as a species on planet earth, reduction of emissions is the most significant contribution of grass pellet combustion. Grass pellet proves to be a better fuel than many other conventional fuels such as coal, wood etc. for domestic as well as industrial purposes.

The future objective will be to perform pyrolysis and gasification; since pellets have proven to have a high calorific value, the pyrolysis of these pellets at different temperatures would lead to the production of biochar of different kinds. Biochars produced by pyrolysis of pellets at high temperatures are effective in the sorption of organic contaminants by increasing the surface area, microporosity and hydrophobicity whereas the biochars obtained at low temperatures are more suitable for removing inorganic/polar organic contaminants by oxygen-containing functional groups, electrostatic attraction, and precipitation. Biochar could replace coal, coconut shell and wood based activated carbon as a low cost sorbent for contaminants and pathogens [10]. Biochar also finds the solution of the latest problem of water purifiers that are not only retaining the impurities from water but also the essential minerals. With proper alterations, both chemically and physically, biochars can be made to selectively adsorb only specific impurities and release the rest of the essential minerals. Not only for the removal of contaminants but biochar can also be used for soil amendment since has it proven to be loaded with nutrients. Hence, pellet biochar finds wide applicability in water remediation and soil amendment.

Gasification of pellets that is, the reaction of pellets with steam or air, which converts it to syngas by careful control of oxygen amount present. Gasification produces a gas fuel that can be used for direct heat generation and electricity production. Gasification being a partial combustion of the solid produces a combustion mixture—solid, liquid and gas, the amounts of which can be controlled by altering temperature, particle size, residence time, pressure, gas composition under which the pellets are treated and the catalyst's identity if one is used [10].

4 Materials and Methods

Pellet production is a combination of sequential steps i.e. pre-processing, drying, grinding, pelletizing, cooling, screening and bagging. Raw material collected in the form of fallen leaves, mowed lawn grass, dried flowers, fruit peels, entire fruit and vegetable pulp. The raw material was dried under sunlight for 8–10 days at an average temperature of 22.5 °C. The dried raw material was then pulverized to a uniform size of 0.5 mm. This pulverized raw material was pre conditioned with starch water which was sourced from the left over liquid after boiling rice. The pelletizer was then set to a temperature range of 70–80 °C. The buildup temperature was regularly checked by the help of an infrared thermometer. The raw material was put into the pelletizer and pellets were obtained. The pellets were allowed to cool and dry naturally in the sun so as to reduce its moisture content to 10%. The pellets were then packed and stored and after a gap of 4–5 days. The pellets were then examined for their shelf life, durability and strength. The following table shows the combination of raw materials used and the amount the amount of starch water.

S. No	Combination of raw material(s)	Ratio	External binder (if any)
1.	Lawn grass + saw dust	4:1	Starch water
2.	Lawn grass + leaves	1:1	
3.	Leaves + saw dust	4:1	
4.	Plant parts other than leaves + leaves	2:3	
5.	Plant parts other than leaves + grass	2:3	
6.	Grass + leaves + saw dust	2:2:1	

5 Conclusion

For optimum pellet production, 10–12% of moisture, 75–80 °C of temperature and 0.5–1 mm of particle size are required. The pellets must be 4 mm in diameter and 45 mm in length. Out of the various compositions of raw materials, pellets composed of lawn grass and saw dust in the ratio 4:1 were found to be the best in terms of strength and durability and hence suitable for pyrolysis and gasification. From the analysis carried out in order to find the calorific value, moisture content, amount of fixed carbon and the percentage composition of macroelements, grass pellets prove to be a better fuel than many other conventional fuels such as coal, wood etc. for domestic as well as industrial purposes.

Acknowledgements Authors are thankful to University of Delhi for providing financial support via Innovation project. Authors are also thankful to Principal, Hans Raj College for providing necessary infrastructure to carry out the research. Authors are indebted to Dr. P. Raman (TERI) for his constant support and guidance.

Conflicts of Interest "The authors declare no conflict of interest."

References

1. Present status of waste management in India and recommendations. Available online: http://civildigital.com/present-status-waste-management-india-recommendations. Accessed on 26 Dec 2016
2. Biochar. Available online: http://www.biochar-international.org/biochar. Accessed 26 Dec 2016)
3. Mohan D, Sarswat A, Sik Y, Pittman CU Jr (2014) Organic and inorganic contaminants removal from water with biochar, a renewable, low cost and sustainable adsorbent—a critical review. Biores Technol 160:191–202
4. Nguyen TH, Cho HH, Poster DL, Ball WP (2007) Evidence for a pore-filling mechanism in the adsorption of aromatic hydrocarbons to a natural wood char. Environ Sci Technol 41:1212–1217
5. Chun Y, Sheng GY, Chiou CT, Xing BS (2004) Compositions and sorptive properties of crop residue-derived chars. Environ Sci Technol 38:4649–4655
6. Bornemann LC, Kookana RS, Welp G (2007) Differential sorption behavior of aromatic hydrocarbons on charcoals prepared at different temperatures from grass and wood. Chemosphere 67:1033–1042
7. Chen B, Chen Z (2009) Sorption of naphthalene and 1-naphthol by biochars of orange peels with different pyrolytic temperatures. Chemosphere 76:127–133
8. Chen B, Yuan M (2011) Enhanced sorption of polycyclic aromatic hydrocarbons by soil amended with biochar. J Soils Sediments 11:62–71
9. Santos LB, Striebeck MV, Crespi MS, Riberio CA, Julio MD (2015) Characterization of biochar of pine pellet. J Therm Anal Calorim 122:21. https://doi.org/10.1007/s10973-015-4740-8
10. Ahmad M, Rajapaksha AU, Lim JE, Zhang M, Bolan N, Mohan D, Vithanage M, Lee SS, Sik Y (2014) Biochar as a sorbent for contaminant management in soil and water: a review. Chemosphere 99:19–33

Impact of Microbial Diversity on Environmental Stability

Meenakshi Sharma and Nidhi Gautam

Don't Repeat the stories invent the new one
You change the conditions I will develop new pathway to survive

Microbes

Abstract Microbial diversity is the most fundamental component of any ecosystem on Earth. Microorganisms are the most abundant and diverse as they comprise the majority of soil biomass. Microbial community plays a crucial role in ecosystem functioning by controlling biogeochemical cycles of elements essential for life, such as carbon (C) and nitrogen (N). They have great potential in energy conversion and regeneration and will likely be an important component of ecosystem response to climate change. The global climate is changing drastically due to; 1-green house gasses such as CH_4-Methane, N_2O-Nitrous oxide, CO_2-Carbon dioxide, Chlorofluorocarbons (CFCs); 2-organic and inorganic pollutants such as heavy metals, ammonia, cyanide, volatile and halogenated compounds. All are released by different industries viz: textile, leather tanning, plastics, pharmaceuticals etc. More than 3.0 million metric tons of toxic chemicals from over 2,000 industries are annually released into the environment. More than 45,360 metric tons (100 million pounds) of toxic compounds are carcinogens. These contaminants not only increased water and agriculture land contamination but also cause various health-related problems. Most of the physio-chemical methods for detoxifying contaminated sites are expensive and release toxins. Therefore, use of Pollution-Eating Bacteria (PEB) viz; *Acinetobacteria*, *Azospirillum*, *Bacillus*, *Proteobacteria*, *Pseudomonas* and *Serratia* is preferred to improve contaminated sites. Unique soil microorganisms also produce beneficial enzymes which degrade pollutants without leaving any toxic intermediates. Various species of *Bacillus*, *Flavobacterium*, *Paenibacillus*, *Pseudomonas*, *Rhodococcus* and *Serratia* cause partial or complete degradation of harmful pesticides like morpholine, methyl parathion and benizimidazole compounds and have dynamic potential to increase crop yield. Microbes produce beneficial enzymes that solubilize highly insoluble nutrients locked up in soil viz: iron (FeIII) and phosphate complexes (Al, Ca and Fe phosphate). These Plant Growth

M. Sharma (✉) · N. Gautam
Department of Botany, Daulat Ram College, University of Delhi, Delhi, India
e-mail: sharmam2feb@gmail.com

V. S. Parmar et al. (eds.), *Green Chemistry in Environmental Sustainability and Chemical Education*, https://doi.org/10.1007/978-981-10-8390-7_8

Promoting Microorganisms (PGPMs) also enhance soil quality. Both biotic and abiotic stresses are responsible for the loss of >40% of agricultural production. Various culturable microbes work as an excellent bio-control agent to protect crops from nutrient stress and plant pathogens such as *Aspergillus*, *Fusarium*, *Pythium*, *Phytophthora*, *Rhizoctonia* and *Sclerotium*. Additionally, *Acinetobacter*, *Arthrobacter*, *Bacillus*, *Marinobacter*, *Microbacterium*, *Panibacillus* and *Pseudomonas* are used for bioremediation and bioaugmentation technology. The present review paper will help in better understanding of microbial ecology and its crucial role in the detoxification mechanism of polluted sites, and environmental stability.

Keywords Biofuel · Climate change · Microbial diversity · Pollution-eating bacteria (PEB) · Plant growth promoting microorganisms (PGPMs)

1 Introduction

Microorganisms are very crucial for the earth to function. They have great potential in energy conversion and regeneration. Microorganisms comprise the majority of soil biomass and diversity [22]. Microorganisms perform many roles in ecosystem functioning viz; crop fertility, detoxifying pollutants, controlling biogeochemical cycling of elements essential for life, such as carbon (C) and nitrogen (N). They have great potential in energy conversion and regeneration and will likely be an important component of ecosystem response to climate change. The global climate is changing drastically due to; 1-green house gasses such as CH_4-Methane, N_2O-Nitrous oxide, CO_2-Carbon dioxide, Chlorofluorocarbons (CFCs); 2-organic and inorganic pollutants such as heavy metals, ammonia, cyanide, volatile and halogenated compounds. Microbes directly affect the soil function and develop balance to soil composition by adapting and purifying the chemical components. Various contaminants not only increased water and agriculture land contamination but also cause various health-related problems. The hazardous pollutants released by the textile, leather tanning, paint, plastic, pharmaceutical, agrochemical, paper, energy, mineral, metal and pulp industries have increased contamination in water, soil and agriculture lands. These contaminants not only affect our environmental and biodiversity adversely, but also cause various health-related problems and lead to pollution borne diseases globally [6, 19]. Various volatile organic compounds (VOCs) and oxides of nitrogen (NOx), hydroxyl radical (OH), the nitrate radical (NO_3), mutagenic polycyclic aromatic hydrocarbons and other complex fine particles are contributing serious health and environmental issues in unendurable manner [5, 44]. According to the 2000 Toxics Release Inventory (TRI) of the U.S. Environmental Protection Agency (EPA), 2.95 million metric tons (6.5 billion pounds) of toxic chemicals from over 2,000 industries are annually released into the environment. The major pollutants (organic and inorganic wastes) are heavy metals, ammonia, cyanide, volatile organic compounds, halogenated organic compounds,

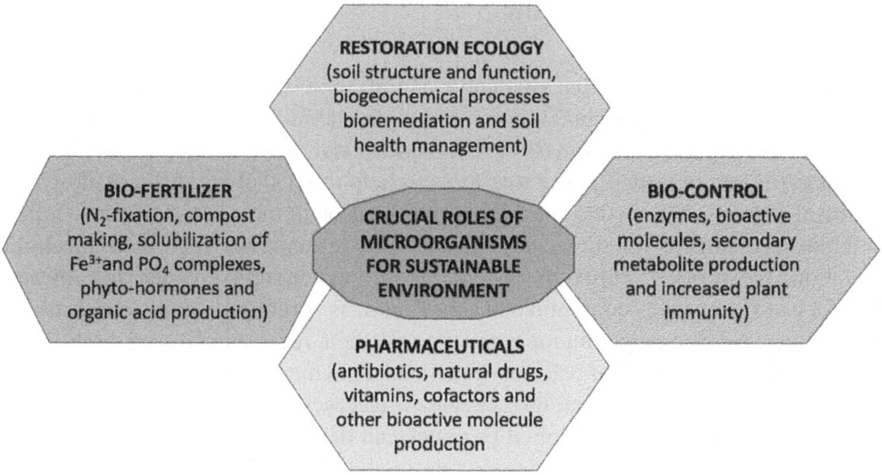

Fig. 1 Major roles of microorganisms for sustainable environment

radionuclides and pharmaceutical substances. More than 45,360 metric tons (100 million pounds) of toxic compounds are carcinogens and around 40% of deaths worldwide are caused due to environmental pollution [12]. Most of the physio-chemical methods used for detoxifying contaminated sites are expensive and release toxins. The present review paper focuses on culturable microbial diversity and their role in agriculture, clean up of polluted sites and environmental restoration (Fig. 1).

Therefore, Pollution-Eating Bacteria (PEB) viz; *Acinetobacteria*, *Arthrobacter*, *Azospirillum*, *Bacillus*, *Marinobacter*, *Panibacillus*, *Proteobacteria*, *Pseudomonas* and *Serratia*, are preferred to be used for bioremediation and bioaugmentation technology to improve contaminated sites. Plant Growth Promoting Microorganisms (PGPMs) also enhance soil quality as they produce beneficial enzymes that solubilize highly insoluble nutrients locked up in soil viz: iron (FeIII) and phosphate complexes (Al, Ca and Fe phosphate) and also work as an excellent bio-control agent to protect crops from nutrient stress and plant pathogens such as *Aspergillus*, *Fusarium*, *Pythium*, *Phytophthora*, *Rhizoctonia* and *Sclerotium*. Various species of *Bacillus*, *Flavobacterium*, *Paenibacillus*, *Pseudomonas*, *Rhodococcus* and *Serratia* that cause partial or complete degradation of harmful pesticides like morpholine, methyl parathion and benizimidazole compounds and have dynamic potential to increase crop yield. It is well known that Microbes form the backbone of every ecosystem and develop a crucial relationship between performance of microbial community and environmental stability.

2 Microbial Detoxification

Several studies proved that bacteria can clean up toxic waste by eating it or by the bacterial enzymes that can break down the toxic chemicals. Some of the microbes such as *Acinetobacteria*, *Bacillus*, *Proteobacteria*, *Pseudomonas*, *Alcanivorax borkumensis* can be added to the tree roots to help them suck up or break down the pollutant present deep inside the agriculture land or dump sites. Microbes follow different mechanisms for the accumulation of toxic materials such as intracellular chelation, biosorption, efflux pump etc. It is a potentially greener and cheaper approach to improve contaminated sites and is preferred over conventional approaches. Additionally, microorganisms and their recombinants are highly efficient for the production of various beneficial enzymes such as oxidoreductases, oxygenases, monooxygenases, dioxygenases, laccases, peroxidases, lipases, cellulases and protease. These microbial enzymes can degrade numerous environmental pollutants without producing any toxic intermediates [24, 47]. The further research in microbial enzymes technology would contribute towards developing advanced bioprocess technology for detoxification of the pollutants and also to obtain novel useful substances. According to the World Health Organization approximately 1.1 billion people of World consume unsafe water. Representative microbial indicators associated with human and animal solid waste are used to detect water pollution level. *Clostridium perfringens*, *Escherichia coli*, *Enterococcus faecalis*, *Enterococcus faecium*, *Enterococcus durans*, *Enterococcus gallinarum*, and *Enterococcus avium* are used as fecal pollution indicators [26].

Various species of *Bacillus*, *Burkholderia*, *Flavobacterium*, *Paenibacillus*, *Pseudomonas*, *Rhodococcus* and *Serratia* cause partial or complete degradation of harmful pesticides like morpholine, methyl parathion and benizimidazole compounds. Different species of *Acinetobacter*, *Arthrobacter*, *Bacillus*, *Panibacillus*, *Pseudomonas*, *Marinobacter* and *Microbacterium* are used in the bioremediation technology (Fig. 1). In addition, some of the microbes produce necessary metabolites required to speed up the rate of degradation of a contaminant and used in bioaugmentation technology [43]. Some novel microbes such as *Pseudomonas putida*, *Pseudomonas aeruginosa*, *Acinetobacter calcoaceticus* and *Alcaligenes odorans* already been reported as crude oil degrading bacteria [1, 18, 28]. These microbes effectively degrade kerosene, gasoline, diesel and engine oil etc. Hazardous wastes affect the biosphere adversely. The footprint of hazardous waste has left highly negative impact on our environment and already contaminated many ocean, lakes, soil and rivers. Microbes by anaerobic digestion can reduce the hazardous effects. Anaerobic digestion is a process by which environmentally hazardous organic wastes from municipal, agricultural and industrial sources may be stabilised and used as renewable energy source. Anaerobic microorganisms like fermentative, syntrophic, acetogenic and methanogenic bacteria decompose complex organic substances into simple, chemically stable compounds like CH_4 and CO_2 [34].

3 Microbes and Soil Fertility

Several novel microbes such as *Azotobacter*, *Azospirillum*, *Bacillus*, *Bradyrhizobium*, *Burkholderia*, *Pseudomonas*, *Rhizobium* and *Thiobacillus* have dynamic potential to increase quality of agriculture yield by fixing atmospheric nitrogen, producing enzymes, plant hormones and organic acids [21, 30]. The successful use of these beneficial microorganisms requires an excellent understanding of ecological interactions that take place between different components of plant-soil-microorganisms [39]. Highly efficient soil microorganisms are one of such microbial technologies for agriculture and environmental stability [41, 42]. Eco-friendly microbes have unique mechanisms to provide natural by products and serve as a good substitute for chemical fertilizers. Microbial-fixed nitrogen serves as a significant source of nitrogen for the agriculture. Microbes are incredibly powerful living thing at the global scale. >50% oxygen and >75% nitrogen are produce by the soil microbes. The microbial enzymes (oxidoreductases, hydrolases, phosphatases and protease etc.) and organic acids (gluconic, formic, citric and acetic acid etc.) solubilize nutrients which are highly insoluble and locked up in soil viz: iron (Fe^{3+}) and phosphates ($AlPO_4$, $CaHPO_4$ Ca_3PO_4 and $FePO_4$) [36, 37]. The plant growth promoting microorganisms (PGPM) not only make these locked nutrients available to the plants, but also enhance the quality of agriculture soil by facilitating resource acquisition, accelerating nutrients cycles, transforming nutrients from unavailable form to available form (carbon, iron, nitrogen and phosphates), modulating phyto-hormones (indole acetic Acid, cytokinins and gibberellins) and modulating the effects of environmental stress, [12, 32, 36]. The beneficial microorganisms, directly or indirectly regulate plant immunity and plant growth through the production of phytohormones, volatile compounds and other metabolites [29, 30]. Microbes like *Azospirillum*, *Bacillus*, *Pseudomonas*, *Rhizobium* and *Thiobacillus* have dynamic potential to increase crop yield up-to 10–40% and fix nitrogen up-to 40–50 kg, when used for 3–4 consecutive years continuously. There is no need of fresh application of biofertilizers further because parental inoculums are sufficient for their growth and multiplication [8]. A unique balance of chemical, physical and biological including microbial components contribute to maintain a healthy ecosystem. Microorganisms are very crucial component of an agricultural soil (Fig. 1). Microbes elicit induced systemic tolerance (IST) in plants against multiple stresses. The presence of significant microbial diversity affects the soil structure, function and makes it healthy in terms of growth enhancement and protection against pests and diseases [4, 39]. Climate change has about 4–9% adverse impact on agriculture each year. As agriculture contributes 15% to India's GDP, climate change presumably causes about 1.5% loss in GDP. Highlighting the impact of climate change on crops, he explained how rice, wheat, maize and sorghum are the worst hit by this phenomenon. By 2030, the global production of major crops like corn, rice, potatoe, wheat are likely to see about 24, 11, 9 and 3% decrease respectively.

4 Microbes as Biocontrol Agents

Both biotic and abiotic stresses are responsible for the loss of >40% of agricultural production worldwide [11, 12, 31]. The use of chemical fertilizers and synthetic fungicides not only cause health hazards but also affect agriculture ecosystems adversely [16, 35]. The value of the global bio-pesticide market is expected to reach $4,556.37 million by 2019, at a compound annual growth rate of 15.30% from 2014 to 2019 (marketsandmarkets.com, 2014). Nowadays, biological control of plant pathogens is of paramount importance as the conventional use of chemical pesticides has been seriously questioned because of environmental and human health hazards. "Biological control is the practice or process by which an undesirable organism is controlled by means of another beneficial organism". Disease are the major biological constraints for crop production and more than sixty pathogens including fungi, bacteria, viruses, mycoplasma and nematodes can infect the crop plants. These pathogens cause soil-born diseases and responsible for severe loss of economically beneficial crops such as *Arachis hypogaea*, *Cicer arietinum*, *Musa acuminate*, *Solanum lycopersicum*, *Triticum aestivum*, *Vigna radiata*, *Zea mays* and other ornamental plants [13, 20, 27]. It is now well established that certain strains of *Bacillus, Pseudomonas*, *Streptomyces* are *Serratia* work as a dominant and highly potential bio-control agents against different plant pathogens such as *Aspergillus*, *Fusarium*, *Pythium*, *Phytopthora*, *Rhizoctonia* and *Sclerotium* [2, 7, 15]. The bio-control abilities of such microbes depend essentially on aggressive root colonization, synthesis of a variety of small molecules, and the production of diffusible or volatile antifungal antibiotics that can inhibit pathogen growth and induction of the systemic resistance in the plant. Several mechanisms such as production of siderophore, antibiotic, HCN, hydrolytic enzymes (β-1, 3-glucanase, CMcellulase, chitinase, lipase, protease, DNase, α-amylase, lecithinase) and volatile compounds for antagonistic activity [2, 7, 12, 20]. The plant growth promoting rhizobacteria (PGPR) are aggressive colonizers of soil and root of the plants and are able to protect plants from infection caused by fungal pathogens such as *Fusarium* spp., *Aspergillus* spp. Plant pathogens are responsible for the loss of 30–40% of agricultural production worldwide, despite the fact that several billion dollar are spent on controlling them with synthetic chemicals and pose serious threat to environment and human health. The control of plant pathogens and disease management in ecofriendly way can only be achieved through an integrative approach by using bio control agents. The better understanding with the use of biocontrol agents can be highly beneficial for sustainable agriculture practices and plant productivity (Fig. 1).

5 Microbial Genes for Crop Improvement

Genetically Modified Microorganisms (GMMs) can produce food additives, amino acids, vitamins, flavours and food supplements, these microbes can also make strong and lighter-weight silk and high quality fibres. Genetic engineering has dramatically expanded the potential of biotechnological methods. Advance technologies now allow us to transfer genes among organisms. Several microbial genes are transferred to plants for the improvement of crop quality such as the beetle resistance gene from *Bacillus thuringiensis* transferred into tomato plants genome. This *Bacillus thuringiensis* gene cry1Ac encodes a protein which is poisonous to certain types of harmful insects. Varieties of Bt-corn and Bt-cotton also contain a gene isolated from soil bacterium *Bacillus thuringiensis* that causes the transgenic corn and cotton to produce an insecticidal protein [3, 10]. Genetically Modified Golden Rice (GMGR) also contains two beta-carotene biosynthesis genes: one is psy gene for phytoene synthase from daffodil (*Narcissus pseudonarcissus*) and second is crtI gene for carotene desaturase from the soil bacterium *Erwinia uredovora* that together result in elevated levels of pro-vitamin A. Therefore, microbial genomics can play significant role for the development of higher quality crops.

6 Microbes as a Source of Next-Generation Biofuel Production

Fuel consumption across the globe has increased drastically with current demand for transportation, power homes, industries, infrastructure and modernization. Around 36–39% of our energy comes from petroleum, 23–29% from natural gas, 18–24% from coal, 11–15% from nuclear electric power and 6–9% from other sources viz. biomass, hydrogen, geothermal, wind and solar. In India total petroleum product consumption increased at $\sim$18–20% in 2016. United States represents less than 5% of the total global population but uses about a quarter of fossil fuel resources of the world. Americans burn around 140 billion gallons of gasoline annually which is increased by 3.0 million barrels per day. Biofuel is very useful and important alternative substitute for fossil fuels. Production of biofuel is increasing at a rapid rate worldwide for energy security, economic development and mitigation of climate change [9, 33, 38]. After plants and yeasts, microbes are the next highly efficient and vital source of biofuel production which play a very important role in green technology. The next-generation biofuel will largely depend on microbes that work as little factories and can be genetically engineered for large scale production [14, 17, 25, 40]. However, synthetic and systems biology is being used in two different processes for biofuels production: first is by using synthetic microbial enzymes to break down biomass into sugars for fuel, and second is the use of novel microbes that produce fuel directly. Biofuel was first proposed in 1970s as the renewable liquid or gases obtained from living organisms.

Diverse scientific projects in different countries are trying to find out wild bacterial strains which have unique properties of the production of sustainable, economical and environment friendly next-generation biofuels including alcohols, isoprenoids, bio-methane (bio-CH_4), biological hydrogen and fatty acid derivatives. Various microbes and their genetically modified strains viz. *Bacillus subtilis*, *Cellvibrio japonicas*, *Clostridium acetobutylicum*, *Corynebacterium*, *Escherichia coli*, *Enterobacter lignolyticus*, *Pseudomonas aeruginosa*, *Pseudomonas putida*, *Rhodosporidium toruloides*, *Rhodotorula glutinis*, *Synechococcus elongates*, *Thermoanaerobacter*, *Yarrowia lipolytica* and *Zymomonas* are already reported as Next-Generation Biofuel producers [17, 23, 40, 45]. Various photosynthetic microbes produce nearly 250 times more fat than plants. Amyris Biotechnologies and LS9 in San Carlos are the two major California-based companies that are working on engineered bacteria for the biodiesel production. Fortunately, the significant genetic manipulations into genetically-tractable model bacterial strain or their synthetic pathways which can produce biofuel at economically relevant yields and rates [14, 46]. The more advance biotechnological tools such as functional genomics, metagenomics, green chemistry, systems biology, synthetic biology, bioinformatics and metabolic engineering can play crucial roles in fulfilling the ever increasing demand of energy by production of advanced biofuels. Therefore, microorganisms may become an ideal platform for the production of biofuel in the future and help to maintain sustainable ecosystems.

7 Future Perspectives

- Inspite of advance technology, we have ambiguous knowledge of the microbiomes (the complete set of microbes in an environment) associated with plants or living freely in different habitats. There are few reports on how the climate change is affecting the microbial diversity and their interactions. It is again unpredictable to calculate the potential of undiscovered microbial genes and their role in sustainable agriculture, industries and ecosystem functions.
- To answer the above challenges, the better understanding of environmental genomics, undiscovered microorganisms, best performing strains, better gene pool discovery and pathway construction for making genetically modified organism can be valuable for human and environmental sustainability.
- The knowledge of microbial diversity and their pathways for the production of enzymes, metabolites, bioactive compounds and antagonistic biomolecules can help in addressing various environmental and agricultural issues in ecofriendly manner.

- Better understanding of interaction of microbes with plants and environment and their applications in microbial biotechnology can help to alleviate various factors responsible for agriculture loss, biogeochemical cycles, climate change and environmental pollution.
- At the time of waste digestion, only 50% of organic compounds undergo biodegradation, further research should be carried out to develop innovative methods for complete degradation of harmful organic wastes.

Conflicts of Interest The authors declare no conflict of interest.

References

1. Abed RMM, Safi NMD, Koster J, Beer D, El-Nahhal Y, Rullkotter J, Garcia-Pichel F (2001) Microbial diversity of a heavily polluted microbial mat and its community changes following degradation of petroleum compounds. Appl Environ Microbiol 68:1674–1683
2. Adesina MF, Lembke A, Costa R, Speksnijder A, Smalla K (2007) Screening of bacterial isolates from various european soils for *in-vitro* antagonistic activity towards *Rhizoctonia solani* and *Fusarium oxysporum*: site-dependent composition and diversity revealed. Soil Biol Biochem 39(11):2818–2828
3. Amarger N (2002) Genetically modified bacteria in agriculture. Biochimie 84(11):1061–1072
4. Arias ME, Jose A, Gonzalez-Perez Francisco J, Gonzalez-Vila Ball AS (2005) Soil health-a new challenge for microbiologists and chemists. Int Microbiol 8:13–21
5. Barbara J, Finlayson-Pitts James N, Pitts J (1997) Tropospheric air pollution: ozone, airborne toxics, polycyclic aromatic hydrocarbons and particles. Science 276:1045–1051
6. Brulle RJ, Pellow DN (2006) Environmental justice: human health and environmental inequalities. Annu Rev Public Health 27:521–536
7. Cao L, Qiu Z, You J, Tan H, Zhou S (2005) Isolation and characterization of endophytic streptomycete antagonists of Fusarium wilt pathogen from surface-sterilized banana roots. FEMS Microbiol Lett 247(2):147–152
8. Cooper R (1959) Bacterial fertilizers in the Soviet Union. Soil Fertil 22:227–233
9. Crutzen PJ, Mosier AR, Smith KA, Winiwarter W (2007) N_2O release from agro-biofuel production negates global warming reduction by replacing fossil fuels. A pioneer on atmospheric chemistry and climate change in the anthropocene, vol 50. Springer, Cham, pp 227–238
10. Dolgin E (2015) Safety boost for GM organisms engineered microbes: kept in check with a synthetic building block. Nature 517:423
11. Glick BR (1995) The enhancement of plant growth by free-living bacteria. Can J Microbiol 41:109–117
12. Glick BR (2012) Plant growth-promoting bacteria: mechanisms and applications. Scientifica 1–15
13. Glick BR, Damir M, Karaturovic P, Newell CA (1995) Novel procedure for rapid isolation of plant growth promoting pseudomonads. Can J Microbiol 41(6):533–536
14. Gronenberg LS, Marcheschi RJ, Liao JC (2013) Next generation biofuel engineering in prokaryotes. Curr Opin Chem Biol 17(3):462–471
15. Hass D, Defago G (2005) Biological control of soil-borne pathogens by fluorescent pseudomonads. Nat Rev Microbiol 3:307–319

16. Horrigan L, Lawrence RS, Walker P (2002) How sustainable agriculture can address the environmental and human health harms of industrial agriculture. Environ Health Perspect 110 (5):445–456
17. Keasling JD, Chou H (2008) Metabolic engineering delivers next-generation biofuels. Nat Biotechnol 26:298–299
18. Lal B, Khanna S (1996) Degradation of crude oil by *Acinetobacter calcoaceticus* and *Alcaligenes odorans*. Appl Biotechnol 81:355–362
19. Law KL, Moret-Ferguson S, Maximenko NA, Proskurowski G, Peocock EE, Hafner J, Reddy CM (2010) Plastic accumulation in the North Atlantic subtropical gyre. Science 329:1185–1188
20. Lecomtea C, Alabouvetteb C, Edel-Hermannc V, Roberta F, Steinberg C (2016) Biological control of ornamental plant diseases caused by *Fusarium oxysporum*: a review. Biol Control 101:17–30
21. Lin TF, Huang HI, Shen FT, Young CC (2006) The protons of gluconic acid are the major factor responsible for the dissolution of tricalcium phosphate by *Burkholderia cepacia* CC-Al74. Bioresour Technol 97:957–960
22. Lorenz P, Eck J (2005) Metagenomics and industrial applications. Nat Rev Microbiol 3:510–516
23. Lu X, Vora H, Khosla C (2008) Overproduction of free fatty acids in *E. coli*: implications for biodiesel production. Metab Eng 10(6):333–339
24. Megharaja M, Ramakrishnana B, Venkateswarlua K, Sethunathand N, Naidu R (2011) Bioremediation approaches for organic pollutants: a critical perspective. Environ Int 37(8): 1362–1375
25. Meng X, Yang J, Xu X, Zhang L, Nie Q, Xian M (2009) Biodiesel production from oleaginous microorganisms. Renew Energy 34:1–5
26. Naidoo S, Olanira AO (2014) Treated wastewater effluent as a source of microbial pollution of surface water resources. Int J Environ Res Public Health 11(1):249–270
27. Navas-Cortes JA, Hau B, Jimenez-Diaz RM (2000) Yield loss in Chickpeas in relation to development of Fusarium wilt epidemics. Phytopathology 90:1269–1278
28. Obayori OS, Adebusoye S, Adewale AO, Oyetibo GO, Oluymi OO, Amokun RA, Ilori MO (2009) Differential degradation of crude oil (Bonny Light) by four *Pseudomonas* strains. J Environ Sci 21:243–248
29. Ortiz-Castro R, Contreras-Cornejo HA, Macias-Rodriguez L, Lopez-Bucio J (2009) The role of microbial signals in plant growth and development. Plant Signal Behav 4(8):701–712
30. Patten CL, Glick BR (2002) Role of *Pseudomonas putida* indoleacetic acid in development of the host plant root system. Appl Environ Microbiol 68:3795–3801
31. Rahi P, Gulayi A (2009) Stress tolerance and genetic variability of phosphate-solubilizing *Pseudomonas fluorescence* from the cold deserts of the trans-himalayas. Microb Ecol 58(2): 425–434
32. Rau N, Mishra V, Sharma M, Das MK, Ahaluwalia K, Sharma RS (2009) Evaluation of functional diversity in rhizobacterial taxa of a wild grass (*Saccharum ravennae*) colonizing abandoned fly ash dumps of Delhi urban ecosystem. Soil Biol Biochem 41:813–821
33. Searchinger T, Heimlich R, Houghton RA et al (2008) Use of US croplands for biofuels increases greenhouse gases through emissions from land-use change. Science 319:1238–1240
34. Shah FA, Mahmood Q, Shah MM, Pervez A, Asad SA (2014) Microbial ecology of anaerobic digesters: the key players of anaerobiosis. Sci World J 42:339–346
35. Sharma M, Laisram N, Singh VP (2010) Screening of a soil bacterium for biological control of pathogenic fungi. J Plant Dev Sci 2:1–4
36. Sharma M, Mishra V, Rau N, Sharma RS (2011) Functionally diverse rhizobacteria of *Saccharum munja* (a native wild grass) colonizing abandoned morrum mine in Aravalli hills (Delhi). Plant Soil 341:447–459
37. Sharma M, Mishra V, Rau N, Sharma RS (2015) Increased iron-stress resilience of maize through inoculation of siderophore-producing *Arthrobacter globiformis* from mine. J Basic Microbiol 55:1–17

38. Singh BK, Richard D, Smith BP, Reay DS (2010) Microorganisms and climate change: terrestrial feedbacks and mitigation options. Nature 8:779–790
39. Singh JS (2015) Microbes: the chief ecological engineers in reinstating equilibrium in degraded ecosystems. Agr Ecosyst Environ 203:80–82
40. Stephanopoulos G (2007) Challenges in engineering microbes for biofuels production. Science 315:801–804
41. Torsvik V, Ovreas L (2002) Microbial diversity and function in soil: from genes to ecosystems. Curr Opin Microbiol 5:240–245
42. Upadhayay SK, Singh JS, Saxena AK, Singh DP (2012) Impact of PGPR inoculation on growth and antioxidant status of wheat under saline conditions. Plant Biol 14:605–611
43. Weyens N, Lelie DV, Artois T, Smeets K, Taghavi S, Newman L, Carleer R, Vangronsveld J (2009) Bioaugmentation with engineered endophytic bacteria improves contaminant fate in phytoremediation. Environ Sci Technol 43(24):9413–9418
44. Wubs ERJ, Wim H, Putten VD, Bosch M, Bezemer TM (2016) Soil inoculation steers restoration of terrestrial ecosystems. Nature 104(2):364–376
45. Yomano LP, York SW, Ingram LO (1998) Isolation and characterization of ethanol-tolerant mutants of *Escherichia coli* KO11 for fuel ethanol production. J Ind Microbiol Biotechnol 20:132–138
46. Zhao Y, Yang J, Qin B, Li Y, Sun Y, Su S, Xian M (2011) Biosynthesis of isoprene in *Escherichia coli* via methylerythritol phosphate (MEP) pathway. Appl Microbiol Biotechnol 90:1915–1922
47. Zwillich TA (2000) Tentative comeback for bioremediation. Science 289(5488):2266–2267

Degradation of DDT, a Pesticide by Mixed Metal Oxides Nanoparticles

Navneet Manav, Vatsala Dwivedi and A. K. Bhagi

Abstract The pesticides use is unavoidable to meet the food production demands of the ever-increasing population. However, these organo-chemicals and their naturally degraded products have tendencies to bio-accumulate and enter into food chain. The toxic nature of these harmful chemicals has become a serious threat to environment in general and to human health in particular. Significantly, high levels of organochlorine pesticides have found to be proliferated into various foodstuffs taken from different regions of India. Average intake of DDT is found to be 48 μg/g, which is quite high. The wide range health hazards and the ability of bioaccumulation of pesticides and their generated waste by-products necessitate the need to degrade these to safer products using suitable economically viable techniques. Mixed metal oxides (MMO) nanoparticles are potential versatile heterogeneous catalysts and may find use in the field of synthetic organic chemistry. The nanostructured binary mixed metal oxides of bivalent metal ion Mg^{2+} with Al^{3+} and Ce^{4+} have been synthesized by sol–gel methods with suitable modifications to increase surface area and to decrease size. The mixed metal oxides have been characterized by several spectroscopic and analytical techniques such as XRD, SEM, TEM. Thereafter the catalysis activity of the nanoparticles was tested on DDT degradation by gas chromatography. Results indicated that nanoparticles are efficient catalyst in remediation of DDT as 90% degradation was achieved in first 24 h.

Keywords Pesticide · Nanoparticles · Degradation · DDT

N. Manav (✉) · A. K. Bhagi
Department of Chemistry, Dyal Singh College, University of Delhi,
New Delhi 110003, Delhi, India
e-mail: navneetmanav@yahoo.co.in

V. Dwivedi
Department of Zoology, Dyal Singh College, University of Delhi,
New Delhi 110003, Delhi, India

V. S. Parmar et al. (eds.), *Green Chemistry in Environmental Sustainability and Chemical Education*, https://doi.org/10.1007/978-981-10-8390-7_9

1 Introduction

Dichlorodiphenyltrichloroethane (DDT) is a stable organic compound introduced in 1942 on a large scale to prevent spread of malaria. Its use was extended to agriculture due to its remarkable insecticidal properties to combat a wide variety of insects. It became one of the most popular pesticide for years because of its stability and persistence of insecticidal action which hazards. Because of its persistent nature it does not degrade rather it tends to bioaccumulate in fatty tissues and is biomagnified. The wide range of human health hazards are from short term impact such as nausea and headache to chronic impacts like cancer, endocrine, reproductive and central nervous system (CNS) disruptions [1] and their ability of bioaccumulation necessitates that excess unused insecticides or their harmful by-products be degraded to safer products using suitable economically viable techniques. Novel catalytic routes for degradation of pesticides and reduction of waste by-products are the key challenges to achieve remediation of agricultural waste water and soil. Nano sized metal oxides are highly active for a large number of reactions that are important both in pollution control and chemical synthesis. In recent years mixed metal oxides (MMO) are being explored as solid catalysts either as active phases or as supports [2]. Bulk MMOs have proven applications in the areas of catalysis and technology because of their versatile chemical and physical properties especially their acid–base and redox catalytic properties [3–10]. Nanomaterials exhibit novel characteristics compared to bulk material due to their special structural variations. Different shapes are known to display different active planes and defects. The size, surface structure and interparticle interaction of nanoparticles leads to unique properties and therefore they have great potential to act as excellent catalysts [11]. Among the metal oxide catalysts, transition metals occupy a predominant place because of their low cost of production, easy regeneration and selective action. They catalyse different types of organic reactions, such as oxidation, isomerisation, dehydration and dehydrogenation. Their catalytic activity may be attributed to the presence of partially filled d-orbital of the transition metal ion and also due to the influence of the oxide ligand field on this partially filled d-orbital. In addition to their ability to form complexes, possibilities of variable oxidation states and high surface area make them versatile catalyst. MMO are combinations of two or more metallic oxides in proportions that may either vary or may be defined by a strict stoichiometry. MMOs obtained in the form of powder or single crystals have a wide spectrum of industrial applications in ceramics, electronics, nuclear research and catalysis. Control over structural and morphological properties is crucial for the performance of MMO nanoparticles in various applications. The high activity of nanosized particles may also be attributed to variation in coordination sites as the crystal size decreases along with the increase of surface sites. Another important aspect of nano-crystalline MMOs is that it provides well arranged oxidising and reducing sites on adjacent positions in a crystal lattice which allow these materials to act as an excellent catalyst for the oxidation and reduction reactions. Some oxides exhibit surface basic behaviour and others as surface acidic behaviour. Mixing two oxides therefore can enhance the catalytic activity manifolds due to synergistic effects. The synthesis of MMOs nanoparticles has attracted attention of synthetic chemists because of the growing concern for use of greener technologies for environmental remediation. In the current study we synthesized combination of MMOs and studied its effect on DDT degradation.

2 Materials and Method

2.1 *Synthesis of Binary Mixed Metal Oxides $MgCeO_3$ and $MgFe_2O_4$*

To synthesize the binary mixed metal oxides, stoichiometric volume of 0.2 M solution of both the corresponding metal salts were mixed in a round bottom flask (Cerium nitrate, Magnesium nitrate, Ferric chloride and Magnesium chloride of Analytical grade were taken). The resulting solution was stirred using a magnetic stirrer. To this solution, while continued stirring a mixture of ammonia and ethanol in 1:9 was added drop wise until the solution attained a particular pH and a gel is formed. The solvent was removed from the gel by rotatory evaporator. The resulting product obtained was then calcined at different temperatures i.e. 400, 500 and 700 °C.

The synthesized metal oxides were characterized by various analytical techniques such as SEM, TEM, and XRD in order to obtain structural information and optimization of calcination temperature of the product. D Bruker—D8 Discover X-ray diffractometer with Cu—Kα radiation generated at 40 kV and 40 mA, was used to analyze X-Ray diffraction patterns for samples.

The average size of nanoparticles in mixed metal oxide were calculated from Scherrer's equation [12–14] $d = K\lambda/\beta \; Cos\theta$ where β is the full width at half maximum (FWHM) of observed diffraction line at its half intensity maximum at an angle 2θ, K is the shape factor which usually have value 0.9 and λ is X-ray wavelength ~ 1.54 nm.

The surface morphology and EDAX was monitored using Scanning Electron Microscope (Model 6610LV, JEOL, Tokyo, Japan). To gain insight of structure and particle size transmission electron microscopy was employed (model TECNAI GT30).

Degradation Kinetic Test:

Test for degradation of DDT was conducted in Teflon capped vials at 2.5 ppm concentration and placed on shaker at 28 °C. Thereafter the sample aliquot was taken at 1h and 24 h. The aliquot taken was centrifuged at 5000 rpm for 10 min to remove the nanoparticles and run on GC system (Agilent Technologies 7890A) to know the final concentration of DDT after 1h and 24 h. DDT was detected at a retention time of 16.9 min while DDD and DDE were detected at 16.2 and 15.5 min respectively. Peak areas of known concentrations of DDT, DDE and DDD were used to draw the standard curves. The curves were used to determine the concentration of DDT, DDD and DDE.

3 Result

Mixed metal oxides such as $MgCeO_3$ and $MgFe_2O_4$ were prepared to test their ability on DDT degradation. The XRD analysis of the prepared catalyst was carried out to characterize the mixed metal oxides formed. The XRD patterns obtained is given in Fig. 1. Powder XRD pattern of $MgCeO_3$ shows it to be monoclinic structure as all peaks can be indexed according to JCPDS file 040641 and $MgFe_2O_4$ as cubic structure as all the peaks can be indexed according to JCPDS file 822424.

The average particle size calculated from XRD is 21.06 nm in $MgFe_2O_4$ and 33.33 nm in $MgCeO_3$. Some of the individual particles are well resolved and their sizes are found to be less than 10 nm in TEM images (Fig. 2), but in general these particles tend to agglomerate as is evident from SEM images (Fig. 3). The EDAX data supported the characterization of mixed metal oxides. Crystallite size of the mixed metal oxides calculated from Scherer's formula is indicated in Table 2.

DDT degradation assay result: The result of DDT degradation on application of different nanoparticles is given in Table 1. It was seen that the area of peak of DDT decreased and DDD (dichlorodiphenyldichloroethane) increased. It was seen that major degradation occurred within the first hour (Table 1). DDT was degraded to DDD, DDE (dichlorodiphenyldichloroethylene) and simpler compounds (Table 2).

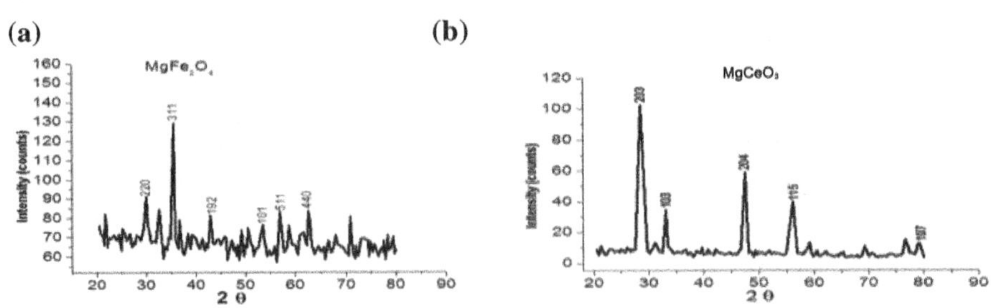

Fig. 1 **a** XRD of $MgFe_2O_4$; **b** XRD of $MgCeO_3$

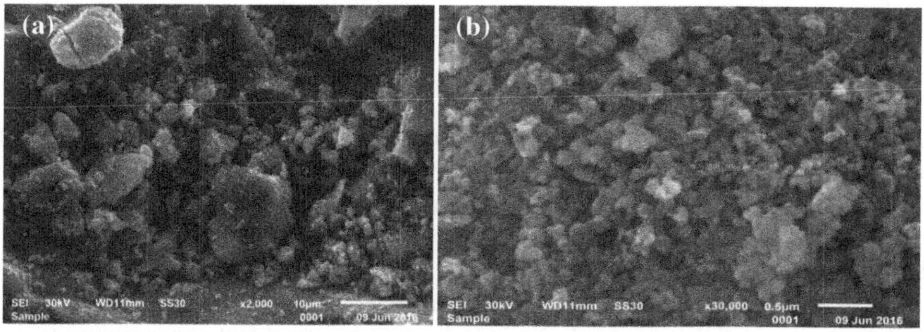

Fig. 2 SEM images of **a** $MgFe_2O_4$; **b** $MgCeO_3$

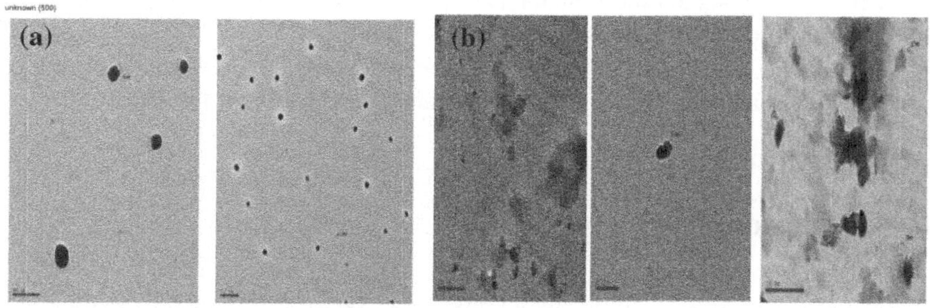

Fig. 3 TEM images of **a** $MgFe_2O_4$; **b** $MgCeO_3$

Table 1 DDT degradation assay at 1 and 24 h by using mixed metal oxide $MgCeO_3$ and $MgFe_2O_4$

Nanoparticle	@1 h			@24 h		
	DDT	DDD	DDE	DDT	DDD	DDE
$MgCeO_3$	0.11	0.29	0.15	0.13	0.31	0.15
$MgFe_2O_4$	0.15	0.30	0.15	0.14	0.31	0.15

Table 2 Crystallite size by Scherrer formula

Composition	FWHM	Peak position	Crystallite size (nm)
$MgCeO_3$	0.2569	28.46	33.33
$MgFe_2O_4$	0.4137	35.42	21.06

3.1 Discussion

The prepared catalyst calcined at different temperatures was subjected to transmission electron microscopy in order to ascertain the nanoparticle size formed. It was observed that size obtained was minimum when the mixed metal oxides were calcined at 700 °C and the particles formed were in varying sizes and in nano regime. It was also observed that there was significant variation in particle size. Particle of size less than 10 nm was also found in TEM images. The scanning electron micrograph (SEM) images of all the synthesized catalysts were measured and it was revealed that the catalysts were devoid of distinct features, pointing to the absence of well-defined morphology. There is no change in the surface morphology when calcined at different temperatures i.e. 400, 500 and 700 °C.

DDT degradation data shows that DDT converts into its lesser toxic compounds in the first hour. From the data it is concluded that although 25% of DDT is degraded to DDD and DDE, but about 75% of DDT has been degraded to other simpler products by these nanomaterials. No significant difference is seen if kept for 24 h.

4 Conclusion

In the present study, a simple and convenient synthesis of mixed metal oxide nanoparticles $MgCeO_3$ and $MgFe_2O_4$ is reported and their average crystal size was found to be 33 and 21 nm respectively.

The degradation kinetic studies of show that in 1 h more than 90% of DDT is degraded to lesser toxic metabolites. Hence the study shows that nanoparticles are effective catalyst in remediation of the insecticide DDT. However, different concentration of nanoparticles needs to be undertaken to understand the optimum concentration and further time kinetics study needs to be done.

Acknowledgements This works was supported from funds under Innovation Project granted by University of Delhi for the project DS305 titled "Degradation of Pesticides using Mixed Metal Oxide (MMO) Nanoparticles as catalysts, Whole cell organisms and Enzymes."

Conflicts of Interest The authors declare no conflict of interest.

References

1. Longnecker MP, Rogan WJ, Lucier G (1997) The human health effects of DDT (dichlorodiphenyltrichloroethane) and PCBS (polychlorinated biphenyls) and an overview of organochlorines in public health. Annu Rev Public Health 18:211–244
2. Gawande MP, Pandey RK, Jayaram RV (2012) Role of Mixed metal oxides in catalysis science-versatile applications in organic synthesis. Catal Sci Technol 2:1113–1125

3. Saqer SM, Kondarides DI, Verykios XE (2011) Catalytic oxidation of toluene over binary mixtures of copper, Mangnese and cerium oxides supported on y-Al_2O_3. Appl Catal B: Environ 103:275–286
4. Monopoli A, Nacci A, Calo V, Ciminale F, Cotugno P, Mangone A, Giammossa LC, Azzone P, Cioffi N (2010) Palladium/zirconium oxide nanocomposite as a highly recyclable catalyst for C-C coupling reactions in water. Molecules 15:4511–4525
5. Gawande MB, Branco PS, Parghi K, Shrikhande JJ, Pandey RK, Ghumann CAA, Bundelski N, Teodoro OMND, Jayaram RV (2011) Synthesis and characterization of versatile MgO-ZrO_2 mixed metal oxide nanoparticles and their applications. Catal Sci Technol 1:1653–1664
6. Gawande MB, Rathi AK, Branco PS, Potewar TM, Velhinho A, Nogueira ID, Tolstogouzov A, Ghumman CAA, Teodoro OMND (2013) Nano-MgO-ZrO_2 mixed metal oxides: characterization by SIMS and application in the reduction of carbonyl compounds and in multicomponent reactions. RSC Adv 3:3611–3617
7. Devade MJ (2015) Synthesis and characterization of CuO-SnO_2 and CuO-SnO_2-Fe_2O_3 mixed metal oxide. J Appl Chem 4:277–283
8. Chen X, Carabineiro SAC, Bastos SST, Tavares PB, Orfao JJM, Pereria MFR, Fihueiredo JL (2014) Catalytic oxidation of ethyl acetate on cerium-containing mixed oxides. Appl Catal A 472:101–112
9. Litt G, Almquist C (2009) An investigation of CuO/Fe_2O_3 catalysts for the gas-phase oxidation of ethanol. Appl Catal B: Environ 90:10–17
10. Stengl V, Hencyh J, Grygar T, Perez R (2015) Chemical degradation of trimethyl phosphate as surrogate for organo-phosphorous pesticides on nanostructured metal oxides. Mater Res Bull 61:259–269
11. Gawande MB, Paula SB, Varma RS (2013) Nano-magnetite (Fe_3O_4) as a support for recyclable catalysts in the development of sustainable methodologies. Chem Soc Rev 42:3371–3393
12. Scherrer P (1918) Bestimmung der Grösse und der inneren Struktur von Kolloidteilchen mittels Röntgenstrahlen. Nachr Ges Wiss Göttingen 26:98–100
13. Bensebaa F, Patrito N, Page YL, Ecuyer PL, Wang D (2004) Tunable platinum–ruthenium nanoparticle properties using microwave synthesis. J Mater Chem 14:3378–3384
14. Kharkwal A, Sharma SN, Jain K, Arora L, Chawla P, Singh AK, Chand S (2013) Polymeric stabilization of hybrid nanocomposites: a comparison between in situ and ex situ-grown $CuInS_2$ poly(3-hexylthiophene)polymer. Colloid Polym Sci 291:2607–2617

Porous Silica Nanoparticles from Rice Husk for the Elimination of Erichrome Black T (EBT) from Laboratory Waste Water

Priti Malhotra, Arti Jain and Ritu Payal

Abstract Water effluents of various chemical laboratories are expected to consist of wide range of organic waste products whose removal is an arduous task and a serious concern with the objective of eradication of all the pollutants from laboratory waste water has been attempted to modify rice husk's surface. The efficacy of the designed adsorbent derived using rice husk and its potential in removing the residual dyes has been evaluated in this research. Both raw silica and functionalized silica obtained from the rice husk have proved to possess greater ability than activated charcoal for the biosorption of dyes present in contaminated water. A comparative account of silica, functionalized silica and activated charcoal acting as biosorbent for a common acidic dye: EBT was studied using FT-IR, TEM, IR spectroscopy techniques for their characterization. It was also observed that adsorption tendency of adsorbent was governed by experimental condition directly. The variation in rates and thermodynamic factor also influences the adsorption efficiency. For the removal of EBT from waste water functionalized silica was revealed to be most efficient.

Keywords Functionalized silica · Activated charcoal · Organic pollutants
Adsorption · Nanosilica

P. Malhotra (✉) · A. Jain
Department of Chemistry, Daulat Ram College, University of Delhi, Delhi 110007, India
e-mail: pritimalhotra21@gmail.com

A. Jain
e-mail: jainarti21@gmail.com

R. Payal
Department of Chemistry, Rajdhani College, University of Delhi, New Delhi 110015, Delhi, India
e-mail: ritupayal.10@gmail.com

V. S. Parmar et al. (eds.), *Green Chemistry in Environmental Sustainability and Chemical Education*, https://doi.org/10.1007/978-981-10-8390-7_10

1 Introduction

Dyes are the major organic pollutant in the current era of modernization and industrialization. Chemical laboratories engaged in cosmetics, leather, textile, paint significantly contain dyes in their waste water. Laboratory waste water belongs to industrial waste waters having lesser amounts of decomposable and chemically stable compounds. The Chemical treatment of textiles involves the usage of various chemicals such as acids, bases, dyes, inorganic and organic auxiliaries, salts, surfactants, along with different volume of water stocks depending on the kind of process, product and equipment [1]. Among various organic dyes EBT is the common azo-dye having various uses such as: indicator in complexometric titration, coloring agent in biological staining and various other industries. It is present as an effluent in such laboratories and undoubtedly contaminates the main stream water bodies to which the effluent water flows. As such the hazards caused by the dyes are a serious problem and needs to be talked through greener technologies. Allowing the dye to degenerate is expected to discharge carcinogenic product, therefore it is essential to remove them from water streams by economically viable and greener methods.

EBT and its degradation products such as naphthaquinone are highly toxic. The literature survey showed that only few papers have raised the removal of EBT [2–6]. Rice plants accumulate silica via crosslinking of the polar silicic acid (H_4SiO_4) which is captivated into insoluble polysilicic acids from the soil, followed by precipitation as amorphous silica and get deposit in peripheral walls of plant cell. In this protocol, rice husk (RH) turns into a bio-refinery plant for high quality amorphous silica [7–9]. Existence of silica in rice husk has been approved since early 19th century [10], from that time various researchers are continuously experimenting on RH to extract silica several [11–18].

Rice husk, an agricultural waste can be recycled and converted into functionalized nanosilica which can feature as a promising adsorbent for dyes. Among various methods employed in the past including physicochemical methods like coagulation, adsorption, precipitation, filtration and oxidation; the process of adsorption has been found to be most effective in decontaminating water from chemical dyes, although methods like reverse osmosis, flocculation, electroflotation and precipitation have also been utilized but adsorption still outshines other techniques because it requires low concentration. Commercially available activated charcoal, silica eucalyptus bark, bottom ash, bagasse, orange, banana peel, hazelnut shell, neem, sawdust, etc. have no doubt shown great deal of adsorption phenomenon in adsorbing dyed on their surface, but nanosilica poses to be spear heading them in adsorption process 2.

2 Materials and Methods

(a) **Acid treatment**: In order to remove numerous soluble particles such as dust, grinded particles, sand, and several other impurities RH was washed excessively with water. Subsequently, RH was dried in an oven for 24 h at ~110 °C. It was ultrasonicated with an HCl solution for ~half an hour with vigorous stirring. On cooling and keeping the solution for 20 h, it was then decanted and washed meticulously with warm water until the solution became acid free, and was designated as RH′. The wet RH was successively dehydrated in an oven at 110 °C for one day (Fig. 1a, b).

(b) **Thermal treatment**: Weighed amounts of RH′ and RH was exposed to thermal treatment to obtain the ash. The samples were heated inside a programmable furnace (Metrex programmable furnace) using different methods. Different rates (2–10 °C/min) as well as temperatures (1000, 700, and 500 °C) were tested. These were designated as ashes (RHAs, Fig. 1c).

(c) **Extraction of silica**: A known weighed amount of RHA was dissolved and heated in 2.5 M sodium hydroxide solution for ~3 h with continuous stirring. The residue was filtered and washed thoroughly with boiling distilled water and a colorless, transparent, and viscous solution was obtained. The solution was allowed to bring at the RT and then a 10 M H_2SO_4 was added dropwise to it with continuous stirring to maintain pH = 2. Afterwards an ammonium hydroxide solution was added to attain a pH = 8.5 and was allowed to stand at RT for ~3 h (Fig. 1d).

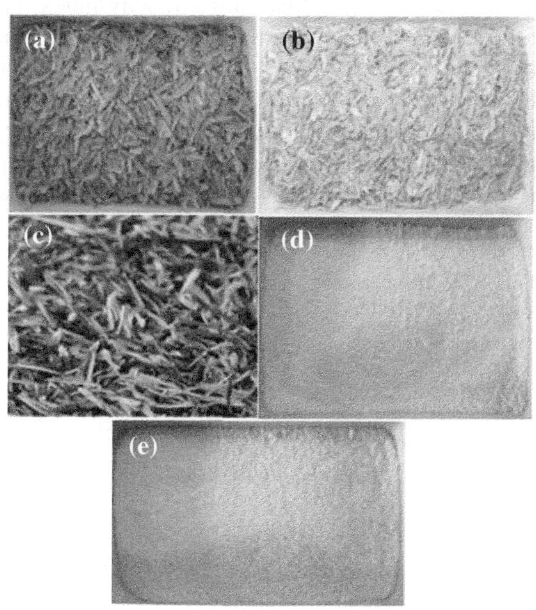

Fig. 1 Illustrates conversion of rice husk into various forms under different conditions: **a** RH, **b** RH′, **c** RHA, **d** silica, and **e** nanosilica

(d) **Preparation of nanosilica**: Nanosilica was synthesized by refluxing the silica with 6 M HCl (acidic solution) for 4 h. The solution was washed repetitively with distilled water to make it acid free and consequently liquefied in NaOH solution by continuous stirring. To ensure the pH = 8 of the solution, H_2SO_4 was added dropwise. The precipitated silica was washed persistently with warm distilled water to free it from excess base.

3 Results and Discussion

3.1 Characterization

Acidic treatment and controlled combustion of RH yielded 25% of RHA. Figure 1d, e revealed both RHA and nanosilica to be white.

Figure 2 showed the FT-IR spectra of obtained silica and nanosilica. A broad band at $\sim 3444\ \text{cm}^{-1}$ is ascribed to the O–H stretching band for Si–OH group. The peak at 1110 cm^{-1} is due to Si–O–Si asymmetric stretching mode while peak at 809 cm^{-1} ascribe the symmetric stretching mode. The band at 469 cm^{-1} describes the Si–O–Si bending mode. A band $\sim 1630\ \text{cm}^{-1}$ is allotted to the bending vibration of water molecules bound to the silica matrix. No band was discovered in the range 2800–3000 cm^{-1}, which means that after controlled combustion and extraction no traces of starting materials were left in the silica.

The assessment of the quantity of unstructured form of silica is important. For resolve this, some specific approaches are described in the literature [19, 20]. One of them documented that 'silica activity index' describes the degree of amorphous form of silica, which is estimated by computing the proportion of silica dissolved in an excess of boiling KOH (4 M) in a 3-min extraction period [21]. This method was investigated for prepared silica and nanosilica, and the results were 99.7 and 99.9% amorphous form, respectively. For further clarification, the XRD spectra were also examined. Figures 3 showed peak for nanosilica present, a broad peak at $2\theta = 22°$ is related with amorphous silica. No sharp peaks were observed which confirms the

Fig. 2 FT-IR spectra of nanosilica and silica

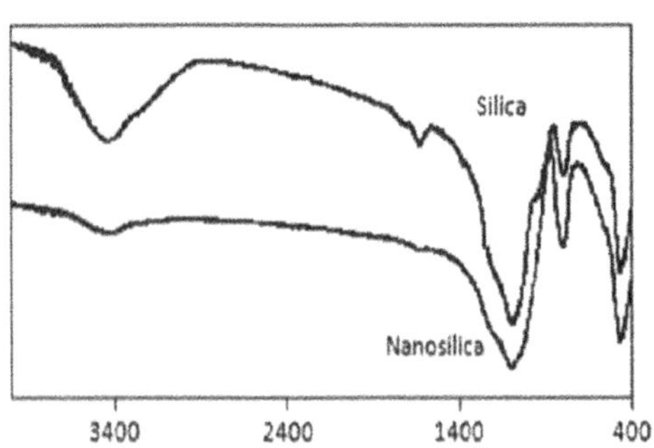

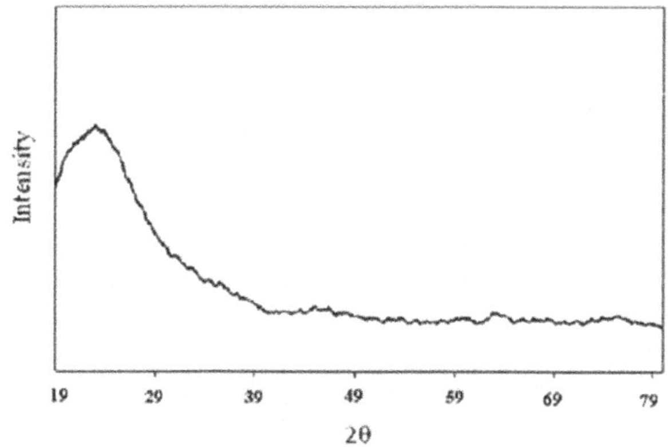

Fig. 3 Powder XRD of nanosilica

lack of any ordered crystalline structure, approving the results from the silica activity index and the non-existence of crystalline phase. Also, there are not at all peaks of carbon, porous or activated, which can be obtained by heating RH at a moderate temperature, this phenomenon occurs because of existence of potassium [21]. However, an acidic treatment of the rick husk removed maximum amount of the potassium, and the ash became free of fixed carbon.

Figure 4 illustrates the TEM micrograph of the nanosilica. The silica particles are spherical in shape with an average particle size of 6 nm.

The metallic constituents have an extensive influence on the quality of silica obtained from RH. It is mainly potassium metal which leads to the surface melting and accelerates crystallization of amorphous silica in addition to carbon fixation in RHA. Moreover, a robust contact amid the metallic ions and silica decreases the surface area. As a result, it's desirable to treat RH with an ideally acidic solution (like HCl in varying concentrations) to efficiently reduce impurities and to get precisely refined silica powder. The chemical reaction amid metals and acid causes a decrease in impurities and then during filtration the reacted metals are seeped from the acidic solution. The color change throughout digestion process is also

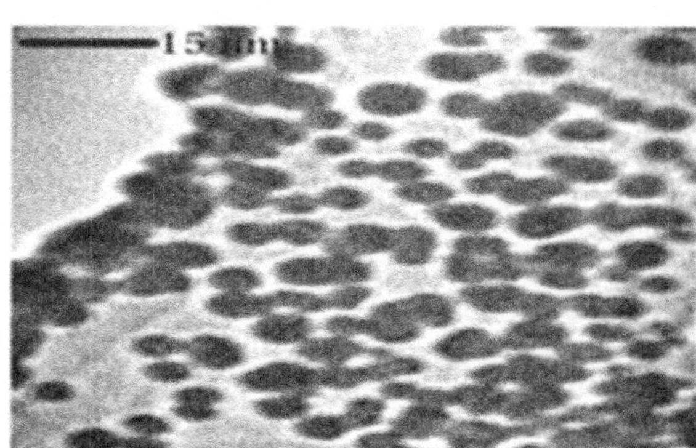

Fig. 4 TEM image of nanosilica

noteworthy. On enhancing the digestion time an initial brown color of RH deviates to yellow or light yellow. Figure 1 show the colors of RH and RH′, respectively on subsequently leaching them with 0.1 M acidic HCl solution for half an hour. The metals perhaps carried out from the volatiles during thermal decomposition. On burning RH gives RHA (ash), which in turn forms highly reactive amorphous silica in controlled conditions. The quality of RHA relied on countless parameters, for eg. time, ashing temperature, soaking time, and rate of heating. The different chemical treatments employed for the elimination of metallic ingredients confirms that percolating with 1 M HCl gives greater pore volume and high surface area. It's noteworthy that a lower combustion temperature (500 °C) gives RHA having greater impurity and it was found and that the crystallization of silica is enhanced on increasing the temperature of combustion up to 1000 °C. Furthermore, it was also established that a combustion temperature of 700 °C yields amorphous silica of high purity and comparatively high surface area.

Quick heating of RH ensures that no oxidation of carbon occurs before the surface melting, which is ensuing large quantity of black particles. Ata heating rate of 10 °C/min, formation of carbon and its oxidation occur even before the dissociation of metal oxides and surface melting. Hence, the affinity of silica to hold carbon is ruled out and a white colored RHA is obtained, which was used for extraction of silica as well as preparation of nanosilica.

Nano sized silica was obtained through precipitation method

$$\text{Silica} + \text{Base} \rightarrow \text{Sodium silicate} + \text{Water}$$

The silica particles were generated from the solution by adding sulfuric acid as catalyst. The acidic condition (around pH 2) reflects the complete precipitation of silica from sodium silicate solution using reaction:

$$\text{Sodium silicate} + \text{Sulphuric acid} \rightarrow \text{Silica} + Na_2SO_4 + H_2O$$

3.2 *Synthesis of Functionalized Nanocomposite*

RHA was boiled in 2.5 M NaOH solution for ~4 h to form the silica from the rice husk ash in the form of sodium silicate. The extracted sodium silicate was then filtered. The solution was further diluted using distilled water and used for single step synthesis of the magnetite–silica nano-composite. In a specific reaction, an aqueous solution containing iron(II) and (III) salts was prepared. This was acquired by adding 2 g of $Fe_2(SO_4)_3$ and 1.39 g of $FeSO_4$ to 20 mL distilled water containing 0.85 mL of HCl solution (12.1 M). In the above solution 10 mL of sodium silicate was added under stirring conditions. During addition, an instant black-brown precipitate was formed, indicating the formation of magnetite particles. The pH of the final solution was maintained at 10. The whole solution was stirred for around 1 h. The particles were separated by a magnetic rod (Fig. 5).

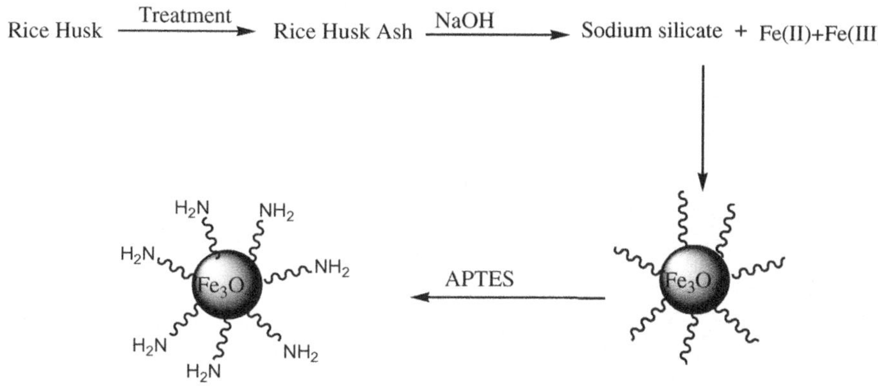

Fig. 5 Complete schematic diagram for the preparation of modified nanocomposite from rice husk

3.3 Purification

The precipitate was washed exhaustively with distilled water and it was purified to separate it from unreacted ions using centrifugation at 8000 rpm for 5 min. The synthesized magnetite–silica composite was desiccated at 60 °C in an oven and used for additional classification as well as surface improvement purposes.

3.4 Surface Modification

The magnetite–silica composite was surface modified by using 3-amino propyl triethoxysilane (APTES). Typically, 1 g of magnetite–silica composite and 8 mL of APTES was added to 2 mL of toluene, with continuously stirring for 3 h. The resultant composite material was washed comprehensively twice with toluene and one time with methanol and purified by centrifugation at the speed of 8000 rpm for 5 min (Fig. 5).

3.5 Removal of Organic Dye EBT from Waste Water: Batch Sorption Studies

All experiments were performed with 10 g/L of sorbent dose in organic dye solutions. Stock solutions (1000 mg/L) were prepared using metal salts and working solutions were prepared with distilled water using appropriate dilutions. All the experiments except effect of pH were measured in the pH range of 6.6–6.8.

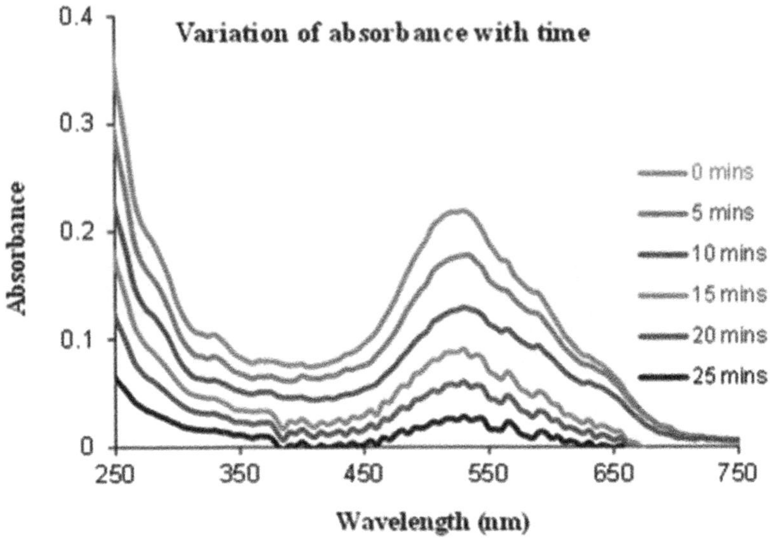

Fig. 6 Variation of absorption w.r.t time

The mixture was ultrasonicated at room temperature. In the end the samples were permitted to centrifuged and then filtered. The supernatant sample was analysed for its absorbance using Motras Scientific atomic absorption spectrophotometer (Figs. 6, 7 and 8).

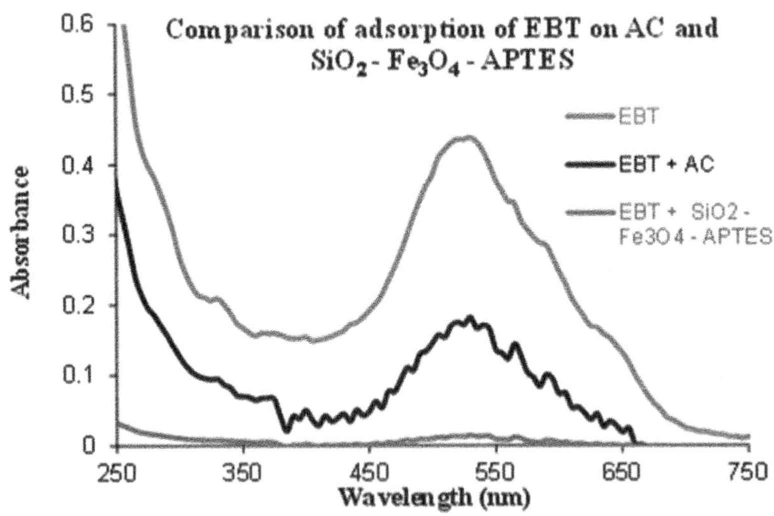

Fig. 7 Adsorption efficiencies of Erichrome Black T

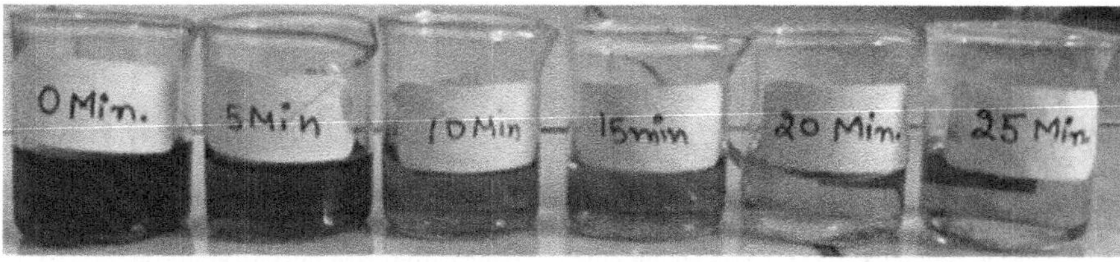

Fig. 8 EBT solutions before and after sorption on aminosilane modified magnetic nanosilica

4 Conclusion

RH on acid treatment followed by thermal combustion underneath controlled conditions results into 22.50% ash which contains 90.469% silica. Under augmented conditions, extremely reactive and purified silica (nanosized silica) with a high surface area was produced along with 99.9% amorphous form. Nanosilica showed strong and excess acidic sites comparative to commercial silica, thereby making it a worthy candidate for adsorption. In view of surplus availability of RH internationally, the investigations shows that eco-friendly protocols can be adopted by rice-growing enterprises to harvest extremely effective silica and to decrease the ecological impact of the usually uncontrolled burning of the industry by-products, specially RH. This is highly significant for agro-based country, which is still the objective of this paper was preparation of an effective biosorbent nanoparticles from a cost effective agricultural by-product (rice husk) and APTES to be used as membrane or nanoparticles for the removal of organic dyes from contaminated water.

Acknowledgements The support from Daulat Ram College and Rajdhani college for providing the laboratory facility to carry out the experimental work is highly appreciated. Financial assistance is provided by "University of Delhi" for innovation project DR-306.

Conflicts of Interest The authors declare no conflict of interest.

References

1. Bilinska L, Bemska J, Ledakowicz S (2012) Inz Ap Chem 51:95 (in polish)
2. Sangale MD, Daptare AS, Sonawane DV (2014) Int J Adv Sci Tech Res 4(4):452–465
3. Stratton WJ, Steehler GA, Pienta NJ, Middlecamp CH (2003) Chemistry in context laboratory manual: applying chemistry to society, 4th edn. McGraw Hill, Boston
4. Dave P, KAur S, Khosla E (2011) S. Indian J Chem Technol 18:53–60
5. Hailei W, Ping L, Guosheng L, Xin L, Jianmimg Y (2010) Rapid Enzym Microb Technol 47:37–43
6. Bedoui A, Ahmadi M, Bensalah N, Gadri A (2009) Chem Eng J 146:98–104
7. Chandra S (2007) Waste materials used in concrete manufacturing. Noyes, Westwood

8. Hwang CL, Wu DS (1989) Properties of cement paste containing rice husk ash. In: Malhotra VM (ed) ACI SP-114: fly ash, silica fume, slag, and natural pozzolans in concrete. American Concrete Institute, Farmington Hills, pp 733–765
9. Lin KM (1975) The study on the manufacture of particle-board made of China fir flakes and hulls. Master thesis, National Chung-Hsing University
10. Martin JI (1938) The desilification of rice hull and a study of the products obtained. Lousiana State University. MS thesis
11. Yalçin N, Sevinç V (2001) Studies on silica obtained from rice husk. Ceram Int 27:219
12. Chandrasekhar S, Pramada PN, Praveen L (2005) Effect of organic acid treatment on the properties of rice husk silica. J Mater Sci 40:6535
13. Chandrasekhar S, Pramada PN, Majeed J (2006) Effect of calcination temperature and heating rate on the optical properties and reactivity of rice husk ash. J Mater Sci 41:7926
14. Chandrasekhar S, Pramada PN, Raghavan P, Satyanarayana KG, Gupta TN (2002) Microsilica from rice husk as a possible substitute for condensed silica fume for high performance concrete. J Mater Sci Lett 21:1245
15. Umeda J, Kondoh K (2010) High-purification of amorphous silica originated from rice husks by combination of polysaccharide hydrolysis and metallic impurities removal. Ind Crop Prod 32:539
16. Pijarn N, Jaroenworaluck A, Sunsaneeyametha W, Stevens R (2010) Synthesis and characterization of nanosized-silica gels formed under controlled conditions. Powder Technol 203:462
17. El-Hosiny FI, Abo-El-Enein SA, Helmy IM, Khalil KA (1997) Effect of thermal treatment of rice husk ash on surface properties of hydrated Portland cement-rice husk ash pastes. J Therm Anal 48:809
18. Mochidzuki K, Sakoda A, Suzuki M, Izumi J, Tomonaga N (2001) Structural behavior of rice husk silica in pressurized hot-water treatment processes. Ind Eng Chem Res 40:5705
19. Payá J, Monzó J, Borrachero MV, Mellado A, Ordoñez LM (2001) Determination of amorphous silica in rice husk ash by a rapid analytical method. Cem Concr Res 31:227
20. Mehta PK (1973) Siliceous ashes and hydraulic cements prepared therefrom. Belgian Patent 802909, July 1973
21. Krishnarao RV, Subrahmanyam J, Kumar TJ (2001) Studies on the formation of black particles in rice husk silica ash. J Eur Ceram Soc 21:99

Synthesis of Spiro[Indene-2,2′-Naphthalene]-4′-Carbonitriles and Spiro [Naphthalene-2,5′-Pyrimidine]-4-carbonitriles via One-pot Three Component Reaction Using Task Specific Ionic Liquid

Pooja Saluja, Garima Khanna, Ankita Chaudhary and Jitender M. Khurana

Abstract A novel and efficient methodology has been developed for synthesizing indane-1,3-dione and pyrimidine based spirocyclic compounds. The synthesis has been achieved via multicomponent reaction of indane-1,3-dione/1.3-dimethyl barbituric acid, variously substituted aromatic aldehydes and cyclohexylidene malononitrile in presence of catalytic amount of TSIL [bmim]OH. Short reaction time, high yields and use of non-chromatographic purification techniques are some of the advantages of this methodology.

Keywords Spiro compounds · Ionic liquid · MCRs · Vinylogous Michael addition · Knoevenagal condensation

1 Introduction

Ionic liquids (ILs) have gained much importance in the field of organic synthesis as they are inflammable, stable over a wide temperature range, usually liquid at room temperature, non-volatile and solvating properties. These have found utility as both reaction medium and catalyst [1]. 1-Butyl-3-methylimidazolium hydroxide ([bmim] OH), a basic task specific ionic liquid (TSIL) has been employed as a catalyst in various organic reactions such as Michael addition [2], Knoevenagel condensation [3], Markovnikov addition [4], Henry Reaction [5], multicomponent reaction [6] etc. One-pot multicomponent reactions (MCRs) can produce highly complex

---

P. Saluja · G. Khanna · J. M. Khurana (✉)
Department of Chemistry, University of Delhi, Delhi 110007, Delhi, India
e-mail: jmkhurana1@yahoo.co.in; jmkhurana@chemistry.du.ac.in

A. Chaudhary
Department of Chemistry, Maitreyi College, Chanakyapuri, New Delhi 110021, India

V. S. Parmar et al. (eds.), *Green Chemistry in Environmental Sustainability and Chemical Education*, https://doi.org/10.1007/978-981-10-8390-7_11

compounds in a single pot without the need of isolation of intermediates which leads to saving of both time and energy [7].

Spiro compounds are the ones in which two carbocycles are joined via single atom. Spirocyclic ketals form the core of various natural products [8], and few spiro motif based drugs have also been investigated [9].

In our aim of developing methodologies for the synthesis of various heterocycles via MCR approach [10–13], we present herein, an efficient methodology for synthesizing of indane-1,3-dione and pyrimidine based heterocycles, involving multicomponent reaction of indane-1,3-dione/1,3-dimethylbarbituric acid, aromatic aldehydes and cyclohexylidene malononitrile using catalytic amount of TSIL [bmim]OH under heating.

2 Results

The reactions of cyclohexylidene malononitrile (I) with aromatic aldehydes viz. 4-cyanobenzaldehyde (IIa)/4-bromobenzaldehyde (IIb)/4-(isopropyl)benzaldehyde (IIc) and indane-1,3-dione (III) were attempted using 20 mol% of basic ionic liquid [bmim]OH at 80 °C. The reactions were completed in 45–60 min and gave the corresponding spiro[indene-2,2′-naphthalene]-4′-carbonitriles namely, 3′-amino-1′-(4-cyanophenyl)-1,3-dioxo-1,3,6′,7′,8′,8a′-hexahydro-1′*H*-spiro[indene-2,2′-naphthalene]-4′-carbonitrile (IVa), 3′-amino-1′-(4-bromophenyl)-1,3-dioxo-1,3,6′,7′,8′,8a′-hexahydro-1′*H*-spiro[indene-2,2′-naphthalene]-4′-carbonitrile (IVb) and 3′-amino-1′-(4-isopropylphenyl)-1,3-dioxo-1,3,6′,7′,8′,8a′-hexahydro-1′*H*-spiro[indene-2,2′-naphthalene]-4′-carbonitrile (IVc) in high yields (Table 1 and Scheme 1).

With the above results in hand, we extended our protocol to synthesize pyrimidine based carbocycles via reaction of cyclohexylidene malononitrile (I), aromatic aldehydes (II), and 1,3-dimethylbarbituric acid (V). Diversely substituted aromatic aldehydes viz. 4-cyanobenzaldehyde (IIa)/4-bromobenzaldehyde (IIb)/4-isopropylbenzaldehyde underwent this three component condensation with cyclohexylidene malononitrile (I) and 1,3-dimethylbarbituric acid (V) to produce 3-amino-1-(4-cyanophenyl)-1′,3′-dimethyl-2′,4′,6′-trioxo-2′,3′,4′,6,6′,7,8,8a-octahydro-1*H*,1′*H*-spiro[naphthalene-2,5′-pyrimidine]-4-carbonitrile (VIa), 3-amino-1-(4-bromophenyl)-1′,3′-dimethyl-2′,4′,6′-trioxo-2′,3′,4′,6,6′,7,8,8a-octahydro-1*H*,1′*H*-spiro[naphthalene-2,5′-pyrimidine]-4-carbonitrile (VIb) and 3-amino-1-(4-isopropylphenyl)-1′,3′-dimethyl-2′,4′,6′-trioxo-2′,3′,4′,6,6′,7,8,8a-octahydro-1*H*,1′*H*-spiro[naphthalene-2,5′-pyrimidine]-4-carbonitrile (VIc) in high yields (Table 1 and Scheme 2).

Table 1 Synthesis of spiro[indene-2,2′-naphthalene]-4′-carbonitriles (IV) and spiro[naphthalene-2,5′-pyrimidine]-4-carbonitriles (VI)

S. No.	Ar(ArCHO) (**I**)	1,3-Dicarbonyl	Product	Time (min)	Yield (%)
1	4-(CN)C_6H_4 (IIa)	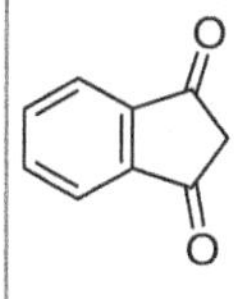	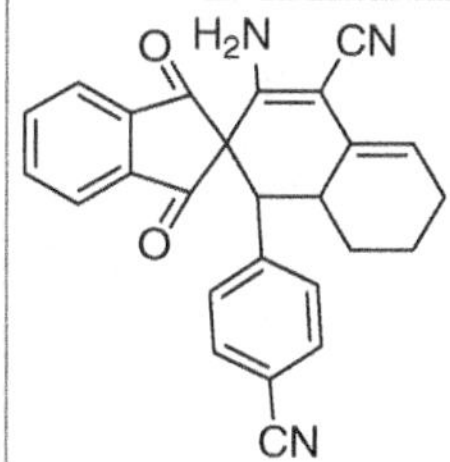 **IVa**	45	86
2	4-BrC_6H_4 (IIb)	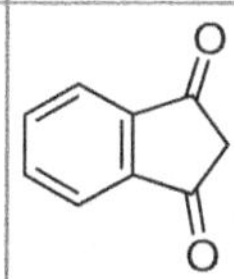	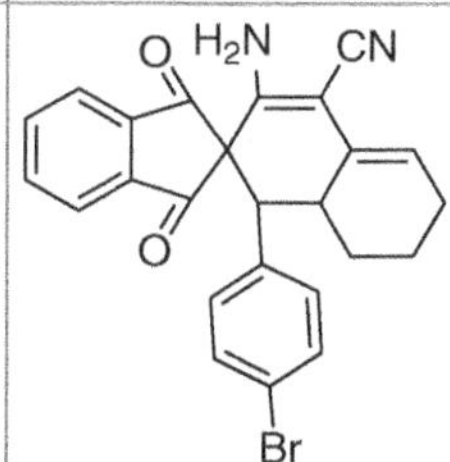 **IVb**	45	84
3	4-$(CH_3)_2CHC_6H_4$ (IIc)	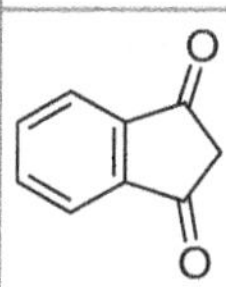	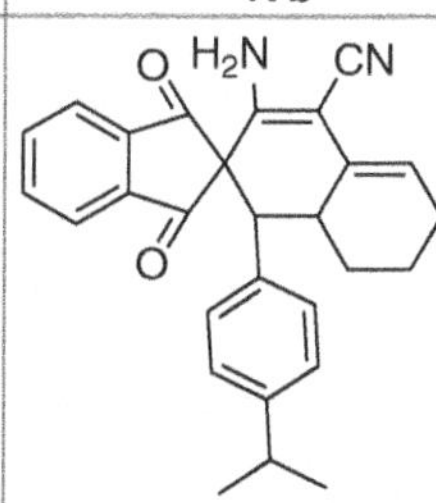 **IVc**	60	81
4	4-(CN)C_6H_4 (IIa)	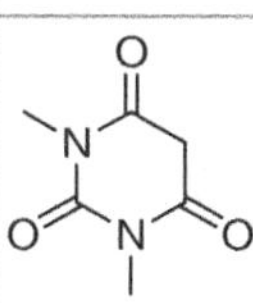	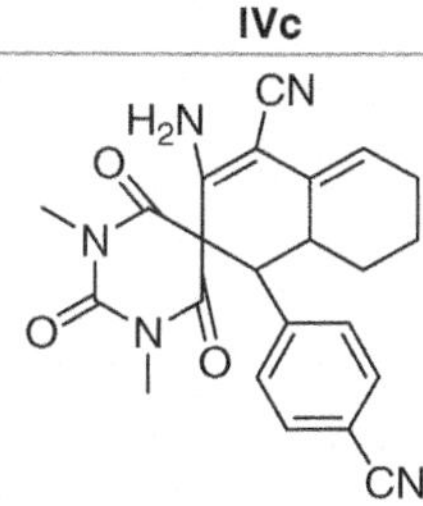 **VIa**	60	87

(continued)

Table 1 (continued)

S. No.	Ar(ArCHO) (**I**)	1,3-Dicarbonyl	Product	Time (min)	Yield (%)
5	4-BrC_6H_4 (IIb)	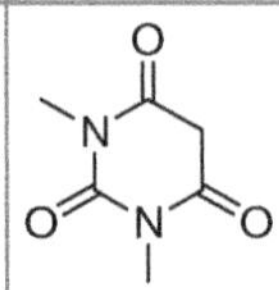	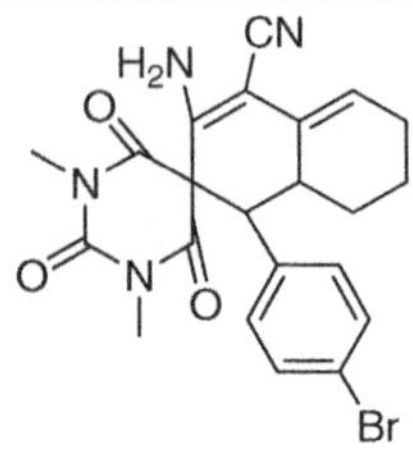 **VIb**	60	85
6	4-$(CH_3)_2CHC_6H_4$ (IIc)	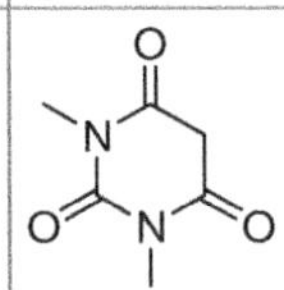	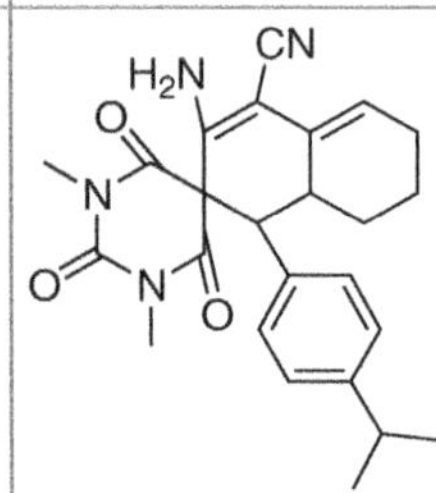 **VIc**	75	82

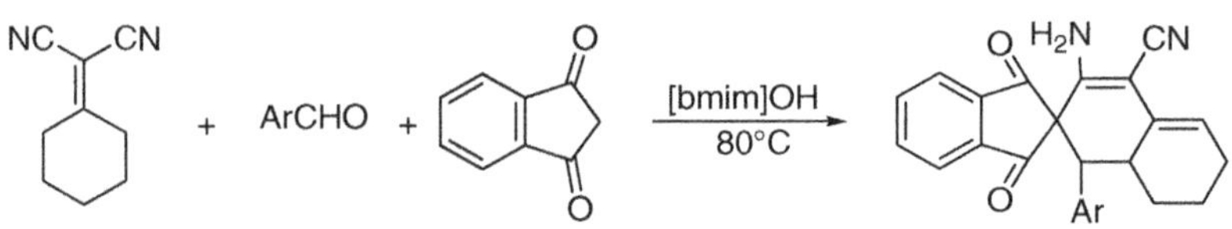

I **IIa-c** **III** **IVa-c**

Scheme 1 Synthesis of spiro[indene-2,2′-naphthalene]-4′-carbonitriles

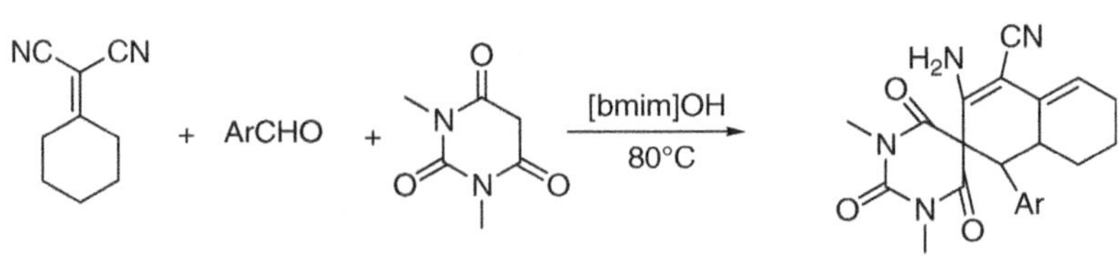

I **IIa-c** **V** **VI a-c**

Scheme 2 Synthesis of spiro[naphthalene-2,5′-pyrimidine]-4-carbonitriles

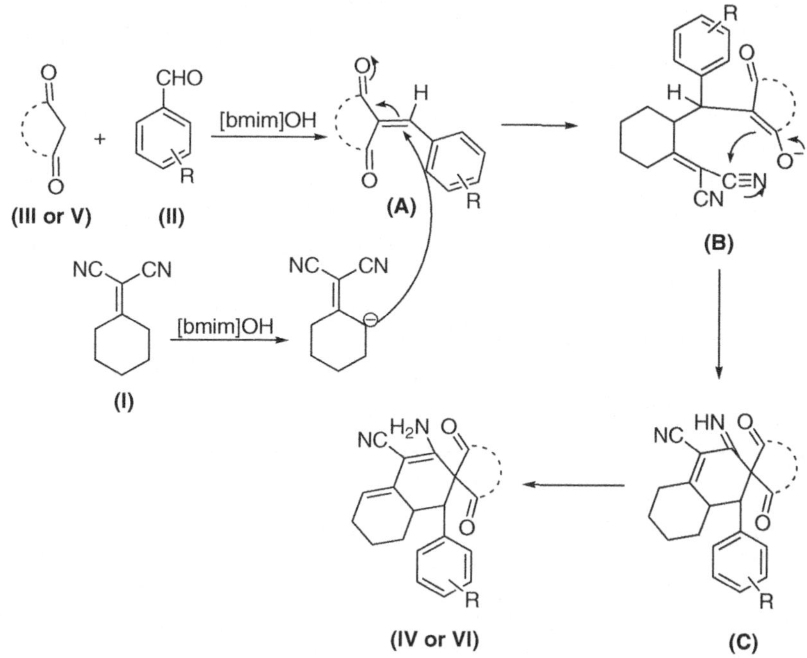

Scheme 3 Plausible mechanism for the synthesis of spirocarbocyclic compounds (**IV/VI**)

3 Discussion

The proposed mechanism for the synthesis of spirocarbocycles (IV and VI) is depicted in Scheme 3. It involves initial Knoevenagel condensation of 1,3-dicarbonyl compound (III/V) with aldehyde (II). The Knoevenagel product (A) so obtained undergoes Michael addition with cyclohexylidene malononitrile carbanion to form an intermediate B. Further, intermediate B undergoes intramolecular nucleophilic addition to give another intermediate C, which upon subsequent isomerization gives desired spirocarbocyclic compounds.

4 Materials and Methods

Buchi melting point 545 apparatus was used for recording the melting point of the synthesized compounds. IR spectra were recorded using KBr pellets on a Perkin Elmer FTIR spectrophotometer, and the values are expressed as ν_{max}/cm^{-1}. NMR spectra were recorded on Jeol JNM ECX-400P instrument. Chemical shift values and coupling constants are reported in δ and hertz (Hz), respectively.

General synthetic procedure for the synthesis of spiro[indene-2,2′-naphthalene]-4′-carbonitriles (IV) and spiro[naphthalene-2,5′-pyrimidine]-4-carbonitriles (VI)

Equimolar amounts of indane-1,3-dione/1,3-dimethylbarbituric acid, aromatic aldehyde and cyclohexylidene malononitrile and 20 mol% of [bmim]OH were taken in a round-bottomed flask and stirred magnetically in an oil bath at 80 °C. The progress of the reaction was monitored using TLC employing petroleum ether: ethyl acetate, (30: 70, v/v) as the eluent. After completion of the reaction as indicated by TLC, the reaction mixture was cooled to room temperature and 5 mL of water was added. The solid precipitated out was filtered using suction pump, washed well with water and then with 1:1 ethanol-water mixture. The structure of the products were confirmed based on spectral analysis.

Spectral Data of Representative compounds:

3′-Amino-1-(4-cyanophenyl)-1,3-dioxo-1,3,6′,7′,8′,8a′-hexahydro-1′*H*-spiro[indene-2,2′-naphthalene]-4′-carbonitrile (IVa)

Yellow solid; m.p.: 247–248 °C (Decomposes); IR (ν_{max}, cm^{-1}) (KBr): 3430, 3338, 3250, 2953, 2235, 2201, 1736, 1698, 1645,1589, 1238; ^{1}H NMR (400 MHz, DMSO-d_6) δ = 7.79–7.71 (m, 3H, ArH), 7.66–7.64 (m, 1H, ArH), 7.45–7.41 (m, 2H, ArH), 7.05–7.02 (m, 1H, ArH), 6.96–6.94 (m, 1H, ArH), 6.21 (s, 2H, NH_2), 5.60 (s, 1H, vinyl), 3.17–3.06 (m, 2H), 2.17–1.96 (m, 2H), 1.63–1.61 (m, 1H), 1.40–1.37 (m, 1H), 1.25–1.22 (m, 1H), 0.78–0.70 (m, 1H); ^{13}C NMR (100 MHz, DMSO-d_6) δ = 199.31, 198.82, 151.06, 142.76, 141.98, 141.80, 136.55, 136.44, 132.12, 131.96, 130.83, 127.78, 123.49, 122.92, 118.26, 122.92, 118.27, 117.61, 117.46, 110.20, 82.46, 63.01, 51.90, 32.72, 27.31, 24.97, 21.57.

3-Amino-1-(4-cyanophenyl)-1′,3′-dimethyl-2′,4′,6′-trioxo-2′,3′,4′,6,6′,7,8,8a-octahydro-1*H*,1′*H*-spiro[naphthalene-2,5′-pyrimidine]-4-carbonitrile (VIa)

White solid; m.p.: 266–268 °C; IR (ν_{max}, cm^{-1}) (KBr): 3451, 3335, 3305, 2230, 2209, 1677, 1638, 1600, 1457, 1379; ^{1}H NMR (400 MHz, DMSO-d_6) δ = 7.79–7.71 (m, 2H, ArH), 7.11–7.05 (m, 2H, ArH), 6.52 (s, 2H, NH_2), 5.56 (s, 1H, vinyl), 3.39 (m, 1H, merged with solvent), 3.18–3.12 (m, 1H), 2.88 (s, 3H, CH_3), 2.85 (s, 3H, CH_3), 2.14–2.01 (m, 2H), 1.60 (brs, 1H), 1.35–1.22 (m, 2H), 0.79–0.74 (m, 1H); ^{13}C NMR (100 MHz, DMSO-d_6) δ = 168.26, 166.73, 150.72, 149.61, 140.81, 132.46, 132.11, 131.50, 130.37, 127.34, 118.45, 117.70, 111.16, 82.17, 60.48, 55.58, 32.02, 28.23, 28.19, 27.26, 24.88, 21.52.

5 Conclusion

In conclusion, we have reported a novel synthetic methodology for the synthesis of spirocyclic compounds using the MCR approach in presence of catalytic amount of [bmim]OH. Indane-1,3-dione/1,3-dimethylbarbituric acid, aromatic aldehyde and cyclohexylidene malononitrile reacted and gave the corresponding spirocycles in good yields. Environment friendly procedure with high atom economy and non-chromatographic purification process are some of the merits of the reported methodology.

Acknowledgements G.K. and P.S. thank CSIR, New Delhi and UGC, New Delhi, respectively, for providing financial assistance in form of junior and senior research fellowship.

Conflicts of Interest Authors declare no conflict of interest.

References

1. (a) Formentín P, Garcia H, Leyva A (2004) Assessment of the suitability of imidazolium ionic liquids as reaction medium for base-catalysed reactions: case of Knoevenagel and Claisen–Schmidt reactions. J Mol Catal A: Chem 214:137–142. https://doi.org/10.1016/j.molcata.2003.05.001; (b) Chen H, Justes DR, Cooks RG (2005) Proton affinities of N-heterocyclic carbene super bases. Org Lett 7:3949–3952. https://doi.org/10.1021/ol0514247
2. Brindaban CR, Jana R, Sowmiah S (2007) An improved procedure for the three-component synthesis of highly substituted pyridines using ionic liquid. J Org Chem 72:3152–3154. https://doi.org/10.1021/jo070015g
3. Yadav LDS, Singh S, Rai VK (2009) A one-pot [Bmim]OH-mediated synthesis of 3-benzamidocoumarins. Tetrahedron Lett 50:2208–2212. https://doi.org/10.1016/j.tetlet.2009.02.166
4. Xu J-M, Liu B-K, Wu W-B, Qian C, Wu Q, Lin XF (2006) Basic ionic liquid as catalysis and reaction medium: a novel and green protocol for the Markovnikov addition of N-heterocycles to vinyl esters, using a task-specific ionic liquid, [bmIm]OH. J Org Chem 71:3991–3993. https://doi.org/10.1021/jo0600914
5. Wu H, Zhang F, Wan Y, Ye L (2008) An efficient protocol for henry reaction using basic ionic liquid [bmim]OH as catalyst and reaction medium. Lett Org Chem 5:209–211. https://doi.org/10.2174/157017808783955817
6. (a) Khurana JM, Chaudhary A (2012) Efficient and green synthesis of 4H-pyrans and 4H-pyrano[2,3-*c*] pyrazoles catalyzed by task-specific ionic liquid [bmim]OH under solvent-free conditions. Green Chem Lett Rev 5:633–638. https://doi.org/10.1080/17518253.2012.691183; (b) Padvi SA, Tayade YA, Wagh YB, Dalal DS (2016) [bmim]OH: an efficient catalyst for the synthesis of mono and bis spirooxindole derivatives in ethanol at room temperature. Chin Chem Lett 27:714–720. https://doi.org/10.1016/j.cclet.2016.01.016
7. Dömling A (2006) Recent developments in isocyanide based multicomponent reactions in applied chemistry. Chem Rev 106:17–89. https://doi.org/10.1021/cr0505728
8. Perron F, Albizati KF (1989) Chemistry of spiroketals. Chem Rev 89:1617–1661. https://doi.org/10.1021/cr00097a015

9. Knox C et al (2011) DrugBank 3.0: a comprehensive resource for 'omics' research on drugs. Nucleic Acids Res 39:D1035–D1041. https://doi.org/10.1093/nar/gkq1126
10. Khanna G, Chaudhary A, Khurana JM (2014) An efficient catalyst-free synthesis of novel benzo[*a*][1,3]oxazino[6,5-*c*]phenazine derivatives *via* one pot four-component domino protocol in water. Tetrahedron Lett 55:6652–6654. https://doi.org/10.1016/j.tetlet.2014.10.067
11. Khurana JM, Chaudhary A, Lumb A, Nand B (2012) An expedient four-component domino protocol for the synthesis of novel benzo[*a*]phenazine annulated heterocycles and their photophysical studies. Green Chem 14:2321. https://doi.org/10.1039/C2GC35644A
12. Khanna G, Aggarwal K, Khurana JM (2015) An efficient and confluent approach for the synthesis of novel 3,4-dihydro-2*H*-naphtho[2,3-*e*] [1,3]oxazine-5,10-dione derivatives by a three component reaction in ionic liquid. RSC Adv 5:46448–46454. https://doi.org/10.1039/c5ra06169e
13. Khanna G, Aggarwal K, Khurana JM (2016) Efficient catalyst-free synthesis of diversified bis (spirooxindoles) *via* one-pot, three-component reaction. Synth Commun 46:1880–1886. https://doi.org/10.1080/00397911.2016.1233437

Recent Progress in the Chemistry of Tri-substituted Triazole via [3 + 2] Cycloaddition of Azide and Double Bond: An Overview

Rajesh Kumar, Manish Kumar, Vipin K. Maikhuri, Navneet Manav, Ashok K. Prasad and Divya Mathur

Abstract Triazoles belong to a class of nitrogen containing compounds which are known for its highly valuable and versatile biological activities. More than 50 years ago, the first synthesis of tri-substituted triazoles was achieved by Huisgen via azide-alkyne cycloaddition. Over the years the remarkable achievement in triazole-based pharmacology promoted ample interest in the discovery of new synthetic methods and catalysts for the synthesis of this miraculous scaffold. Recent tri-substituted triazoles development based on various azides and double bonds have been summarized in this report.

Keywords Catalyst · Cycloaddition · Regioselective · Tri-substituted triazoles

1 Introduction

After the development of click chemistry [1], the 1,2,3-triazole-containing five membered ring emerged as charismatic molecules endowed with a wide spectrum of pharmacological applications [2–4] due to their anticancer, anti-HIV, antitu-berculosis, antifungal, antibacterial antiviral, antiallergic, anti-HIV and antimicro-bial activities etc. Some representative biologically active triazoles are shown in Fig. 1. In addition, they have also achieved new milestones in other areas such as

R. Kumar (✉) · M. Kumar · V. K. Maikhuri · A. K. Prasad
Bioorganic Laboratory, Department of Chemistry, University of Delhi,
New Delhi 110007, Delhi, India
e-mail: rajeshenzyme@gmail.com

N. Manav
Dyal Singh College, University of Delhi, New Delhi 110003, Delhi, India

D. Mathur
Daulat Ram College, University of Delhi, Delhi 110007, India

V. S. Parmar et al. (eds.), *Green Chemistry in Environmental Sustainability and Chemical Education*, https://doi.org/10.1007/978-981-10-8390-7_12

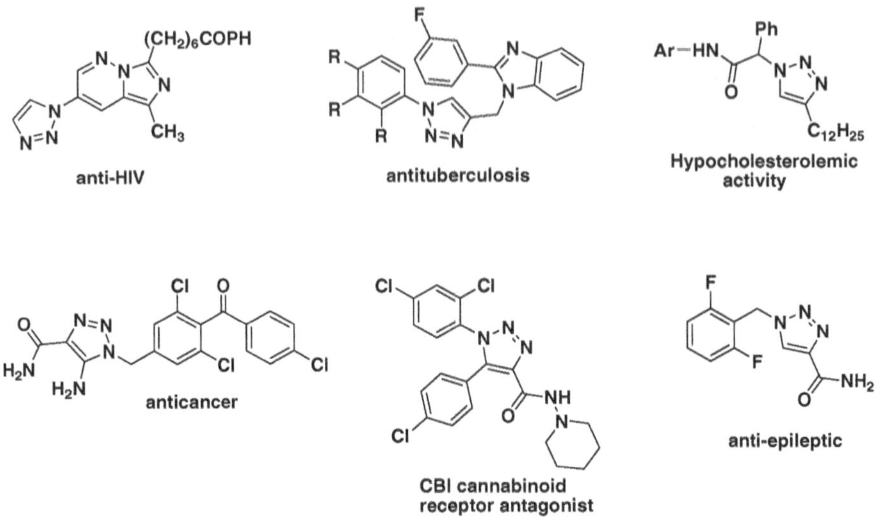

Fig. 1 Representative examples of bioactive 1,2,3-triazoles

drug discovery [5], polymers [6], materials science [7], chemical synthesis [8], agricultural [9], industrial [10], supramolecular chemistry [11], bioconjugations [12] and pesticides [13]. Industrial applications include anti-corrosives [10], dyes, photostabilizers, photographic materials [14].

The first attempt at the synthesis of 1,2,3-triazoles was done successfully by Huisgen [15] in 1963 using 1,3-dipolar cycloaddition reaction between azides and alkynes. But it had inadequate utilization in general synthesis because of poor regioselectivity and high temperature. In order to overcome these limitations, Sharpless, Meldal and other groups developed copper and ruthenium-catalyzed azide-alkyne cycloaddition reactions to procure triazoles with different regioselectivity [16]. Copper-catalyzed reaction is selective for the synthesis of 1,4-disubstituted triazoles and uses terminal alkynes. Ruthenium catalysis leads to 1,5-disubstituted triazoles and can also use internal alkyne (Eq. (I), Scheme 1).

A new class of heterocyclic compounds that have joined the league of pharmacologically important triazoles are 1,4,5-trisubstituted 1,2,3-triazoles [17]. Fully decorated 1,2,3-triazoles have been achieved through various synthetic routes but have not been explored exhaustively [18]. Cycloaddition of azides and internal alkynes have been utilized to accomplish 1,4,5-trisubstituted 1,2,3-triazoles [19]. Recently, Tsai et al. [20] and Majireck and Weinreb [21]. have reported the synthesis of 4,5-disubstituted *2H*- 1,2,3-triazoles by cycloaddition of alkyl azides or metal azides with internal alkynes. The main drawback of this method is the low yielding synthesis of the alkyne reactant, especially when an electron donating moiety is attached to the alkyne terminus.

As an alternative method to azide-alkyne cycloaddition, olefins have been used instead of alkynes to access 1,2,3-triazoles. Major work on azide-olefin

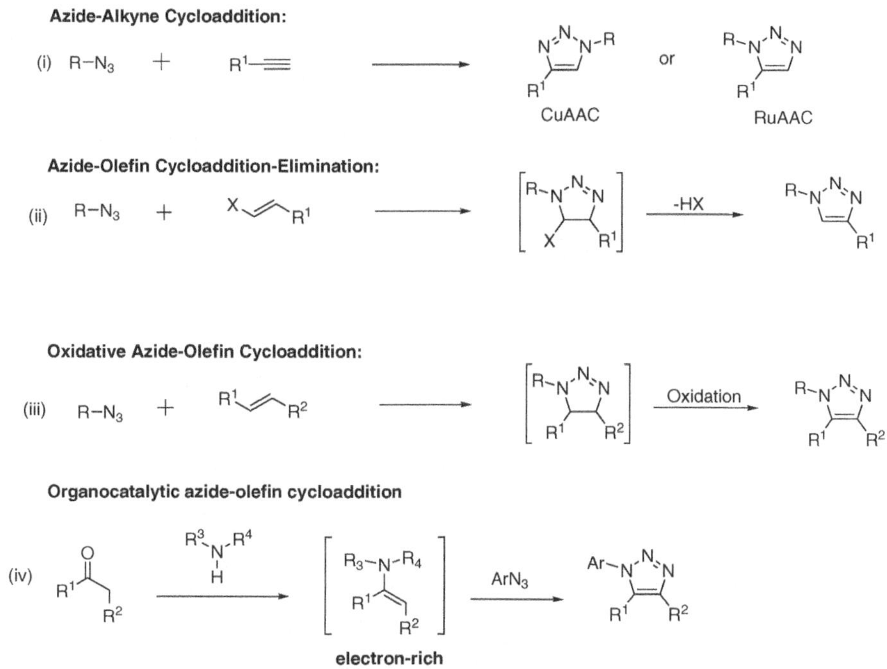

Scheme 1 Azide-olefin cycloaddition reactions

cycloaddition reaction was performed by Huisgen and L'abbe [22] with electron deficient olefins and azides to form triazoles but these triazolines were unstable and they generally kibble into various products depending on the reaction circumstances [23]. To circumvent this, methodologies were developed to aromatize this unstable triazolines into stable triazoles. In this context, olefins containing leaving groups were employed which could afford the triazoles by cycloaddition followed by elimination reaction (Eq. (ii), Scheme 1). Olefins such as nitro olefins [24], vinylacetate [25], alkoxy [26], vinyl sulfones [27] and push-pull olefins [28] were subjected to cycloaddition-elimination reactions with different azides to afford the 1,2,3-triazoles (Fig. 2). Quiclet-Sire and Zard [29] and Sengupta et al. [30] reported the synthesis of 4,5-disubstituted- 1H-1,2,3-triazoles and 2H-1,2,3-triazoles from nitroalkenes via this approach.

Cycloaddition reaction of olefin without any leaving group and azide is another potential method which is yet to be studied extensively in the triazole chemistry. The resulting triazolines thus formed having no leaving group can be aromatized into stable triazoles by its concomitant oxidation (Eq. (iii), Scheme 1). Oxidative azide-olefin cycloaddition reaction (OAOC) has been reported by some research groups and various metal catalysts such as CuI [31], $Ce(OTf)_3$ [32] CuO [33], Cu $(OTf)_2$ [34] $Cu(OAc)_2$ [35] CuBr [36], CuO [33, 37, 38] and Fe_2O_3 nanoparticles [39] have been employed using organic azides with electron-deficient olefins (Fig. 3).

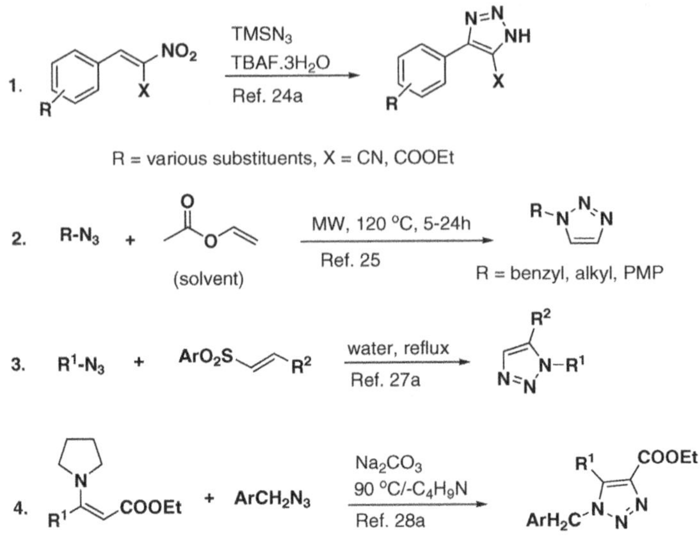

Fig. 2 Representive example of azide olefin cycloaddition elimination reaction

Yao group has demonstrated Cu (I) promoted oxidative cycloaddition reaction of electron deficient internal and terminal olefins with azides under molecular oxygen and basic condition. Xie et al. have presented the $Ce(OTf)_3$ catalyzed oxidative cycloaddition reaction of benzylazides and chalcones. In addition, electron-deficient olefins such as 1,4-naphthaquinone [40] and 1,4-benzoquinone [41] have also been subjected to oxidative cycloaddition reaction with numerous azides (Fig. 4).

In addition, "organo-click" reactions were developed [42] in which organocatalysts were employed instead of metal catalysts. A major disadvantage of copper catalysis is their cytotoxicity, because of which they are not preferred for the labeling of biomolecules in living cells [43]. Moreover, problems such as hepatitis, Alzheimer's disease and neurological disorders can also be caused by excessive intake. Therefore, it is imperative to build up metal-free protocols to synthesize 1,2,3-triazoles. In response to this overwhelming demand, the K_2CO_3-promoted OAOC of chalcones and azides in aqueous medium was recently reported [44].

Tripathi et al. [45]. has reported OAOC of chalcones with various azides in the presence of tetrabutylammonium hydrogen sulphate (TBAHS) as a catalyst in *N,N*-dimethylformamide (DMF) as a solvent in good yield. Gangaprasad's research group has reported metal-free, TEMPO-promoted, oxidative cycloaddition of azides with various electron-deficient olefins in water as a medium to construct a diverse array of trisubstituted 1,2,3-triazoles (Fig. 5) [44].

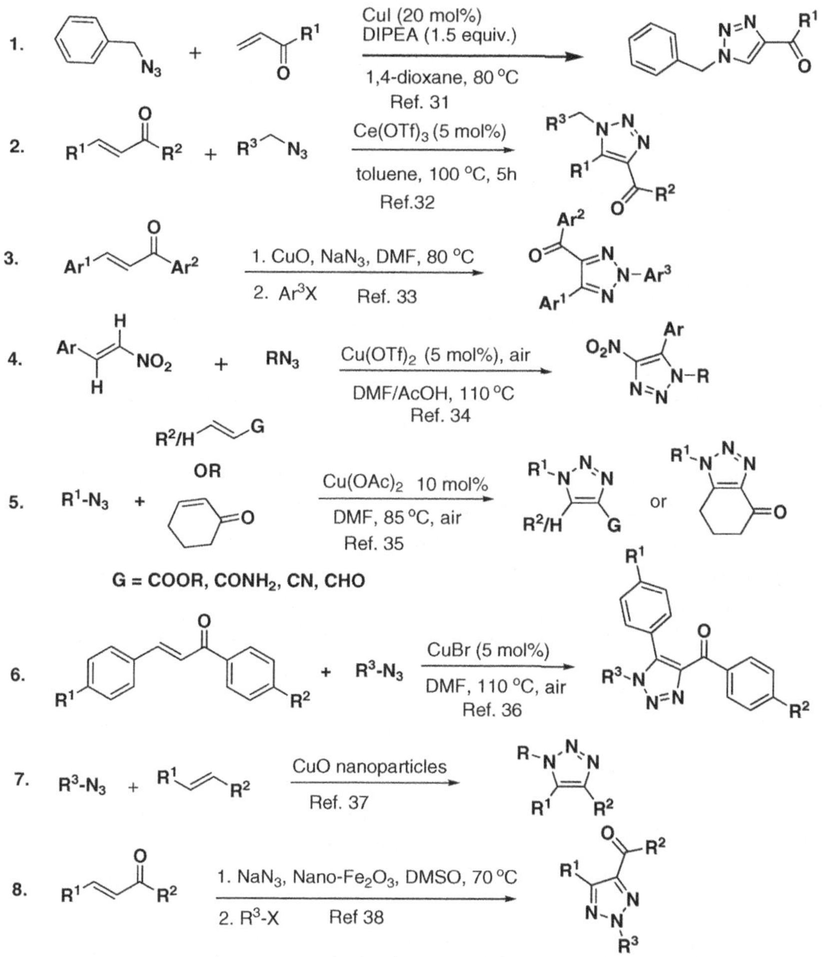

Fig. 3 Representive example of oxidative azide-olefin cycloaddition reaction

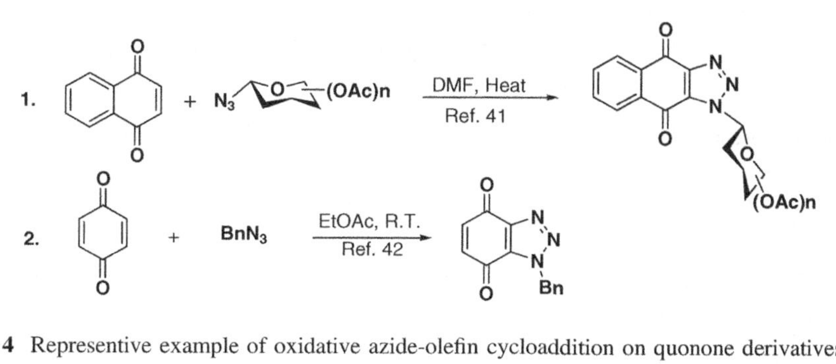

Fig. 4 Representive example of oxidative azide-olefin cycloaddition on quonone derivatives

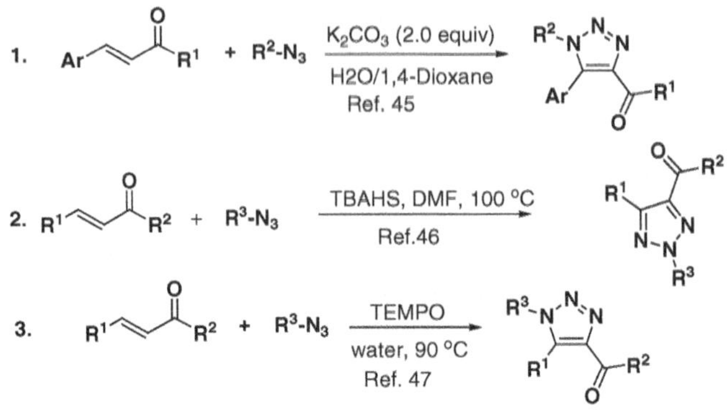

Fig. 5 Representive example of transition metal free oxidative azide-olefin cycloaddition reaction

In addition, "organo-click" reactions were developed to prepare triazoles [46]. "Organo-click" reactions encapsulate the eliminative azide-olefin cycloaddition (EAOC) of enamines (electron-rich olefins) prepared in situ from various activated carbonyl compounds and nitriles that with organic azides catalyzed by amines (Eq. (iv), Scheme 1). In addition to enamine-mediated organo-click reactions, an enolate-mediated organo-click reaction has also been reported (Fig. 6) [47, 48].

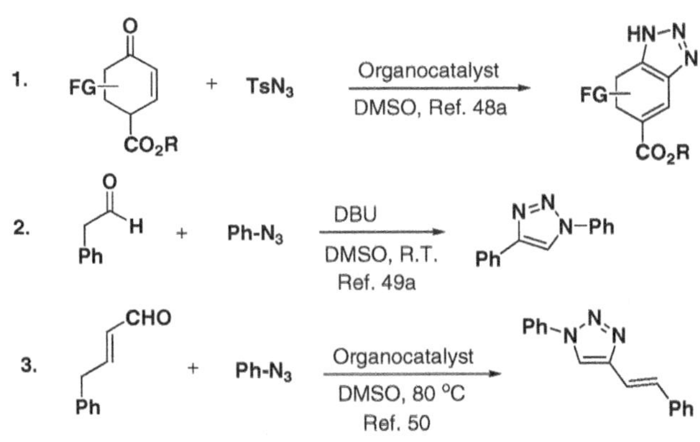

Fig. 6 Representive example of eliminative azide-olefin cycloaddition reaction

2 Conclusion

Triazoles because of their extraordinary binding ability to various proteins and enzymes are gaining attention of various research groups. Fully substituted 1,4,5-trisubstituted 1,2,3-triazoles have gained much interest due to their prominent biological activities. Azide-alkyne cycloaddition is a classic method for the synthesis of 1,2,3-triazoles. Various alternative procedures have also been explored over the years which also include strategies based on azides and double bonds. However, much research is yet to be done for establishing convenient procedures which should be high yielding, regio- and enantiospecific and utilizing mild catalysts and reaction conditions.

Acknowledgements The authors thank University of Delhi for providing financial support under DU-DST purse grant.

References

1. Kolb HC, Finn MG, Sharpless KB (2001) Click chemistry: diverse chemical function from a few good reactions. Angew Chem Int Ed 40:2004–2021
2. Agalave SG, Maujan SR, Pore VS (2011) Click chemistry: 1,2,3-triazoles as pharmacophores. Chem Asian J 6:2696–2718
3. (a) Shu H, Izenwasser S, Wade D, Stevens ED, Trudell ML (2009) Synthesis and CB1 cannabinoid receptor affinity of 4-alkoxycarbonyl-1,5-diaryl-1,2,3-triazoles. Bioorg Med Chem Lett 19:891–893; (b) Cheng H, Wan J, Lin MI, Liu Y, Lu X, Liu J, Xu Y, Chen J, Tu Z, Cheng YSE, Ding K (2012) Design, synthesis, and in vitro biological evaluation of 1H-1,2,3-triazole-4-carboxamide derivatives as new anti-influenza. A agents targeting virus nucleoprotein. J Med Chem 55:2144–2153; (c) Taddei M, Ferrini S, Giannotti L, Corsi M, Manetti F, Giannini, G, Vesci L, Milazzo FM, Alloatti D, Guglielmi MB, Castorina M, Cervoni ML, Barbarino M, Fodera R, Carollo V, Pisano C, Armaroli S, Cabri W (2014) Synthesis and evaluation of new Hsp90 inhibitors based on a 1,4,5-trisubstituted 1,2,3-triazole scaffold. J Med Chem 57:2258–2274; (d) Demchuk DV, Samet AV, Chernysheva N B, Ushkarov VI, Stashina GA, Konyushkin LD, Raihstat MM, Firgang SI, Philchenkov AA, Zavelevich MP, Kuiava LM, Chekhun VF, Blokhin DY, Kiselyov AS, Semenova MN, Semenov VV (2014) Synthesis and antiproliferative activity of conformationally restricted 1,2,3-triazole analogues of combretastatins in the sea urchin embryo model and against human cancer cell lines. Bioorg Med Chem 22:738–755; (e) Gilchrist J, Dutton S, Bustamante MD, McPherson A, Olivares N, Kalia J, Escayg A, Bosmans F (2014) Nav1.1 modulation by a novel triazole compound attenuates epileptic seizures in rodents. ACS Chem Biol 9:1204–1212
4. (a) Alvarez R, Velazquez S et al (1994) 1,2,3-triazole-[2,5-bis-O-(tert-butyldimethylsilyl)-.beta.-D-ribofuranosyl]-3′-spiro-5″-(4″-amino-1″,2″-oxathiole2″,2″-dioxide) (TSAO) analogs: synthesis and anti-HIV-1 activity. J Med Chem 37:4185–4194; (b) Tam A, Arnold U, Soellner MB, Raines RT (2007) Protein prosthesis: 1,5-disubstituted[1,2,3]triazoles as cis-peptide bond surrogates. J Am Chem Soc 129:12670–12671; (c) Carta A, Palomba M, Paglietti G, Molicotti P, Paglietti B, Cannas S, Zanetti S (2007) [1,2,3] triazolo[4,5-h] quinolones a new class of potent antitubercular agents against multidrug resistant Mycobacterium tuberculosis strains. Bioorg Med Chem Lett 17:4791–4794

5. (a) Kolb HC, Sharpless KB (2003) The growing impact of click chemistry on drug discovery. Drug Discovery Today 8:1128–1137; (b) Thirumurugan P, Matosiuk D, Jozwiak K (2013) Click chemistry for drug development and diverse chemical–biology applications. Chem Rev 113:4905–4979
6. (a) Lutz J.-F (2007) 1,3-dipolar cycloadditions of azides and alkynes: a universal ligation tool in polymer and materials science. Angew Chem Int Ed 46:1018–1025; (b) Kempe K, Krieg A, Becer CR, Schubert US (2012) "Clicking" on/with polymers: a rapidly expanding field for the straightforward preparation of novel macromolecular architectures. Chem Soc Rev 41:176–191; (d) Beghdadi S, Miladi IA, Drockenmuller E et al (2012) Synthesis and polymerization of C-vinyl- and N-vinyl-1,2,3-triazoles. Polym Chem 3:1680–1692
7. (a) Xi W, Scott TF, Kloxin CJ, Bowman CN (2014) Click chemistry in materials science. Adv Funct Mater 24:2572–2590; (b) Binder WH, Sachsenhofer R (2008) Click' chemistry in polymer and material science: an update. Macromol Rapid Commun 29:952–981
8. (a) Wacharasindhu S, Mansour TS et al (2009) Oxidative palladium catalysis in SNAr reactions leading to heteroaryl ethers from pyridotriazol-1-yloxy heterocycles with aryl boronic acids. J Am Chem Soc 131:4174–4175; (b) Chattopadhyay B, Gevorgyan V (2012) Transition-metal-catalyzed denitrogenative transannulation: converting triazoles into other heterocyclic systems. Angew Chem Int Ed 51:862–872; (c) Chattopadhyay B, Gevorgyan V (2012) Rhodium-catalyzed N–H insertion of pyridyl carbenes derived from pyridotriazoles: a general and efficient approach to 2-picolylamines and imidazo[1,2-a]pyridines. Angew Chem 124:886–896; (d) Seo B, Jeon WH, Kim J, Kim S, Lee PH (2015) Synthesis of fluorenes via tandem copper-catalyzed [3 + 2] cycloaddition and rhodium-catalyzed denitrogenative cyclization in a 5-exo mode from 2-ethynylbiaryls and N-sulfonyl azides in one pot. J Org Chem 80:722–732; (e) Kim C.-E, Park Y, Park S, Lee PH (2015) Diastereoselective synthesis of tetrahydrofurano- and tetrahydropyrano-dihydropyrroles containing N,O-acetal moieties via rhodium-catalyzed transannulation of N-sulfonyl-1,2,3-triazoles with Oxacycloalkenes. Adv Synth Catal 357:210–220
9. Fan W. In comprehensive heterocyclic chemistry II; Katritzky AR, Rees CW, Scriven EFV (eds) (1996) Pergamon, Oxford vol 4, pp 1–126
10. Umawatari T, Kondo S, Jpn Kokai, Tokkyo Koho (2002) Fluoro-contained 1H-1,2,3-triazaazole compounds, preparation method thereof and their use. JP 2002105672, Chem Abstr 136:298315
11. (a) Fahrenbach AC, Stoddart JF (2011) Reactions under the click chemistry philosophy employed in supramolecular and mechanostereochemical systems. Chem Asian J 6:2660–2669; (b) Schulze B, Schubert US (2014) Beyond click chemistry-supramolecular interactions of 1,2,3-triazoles. Chem Soc Rev 43: 2522–2571
12. (a) El-Sagheer AH, Brown T (2010) Click chemistry with DNA. Chem Soc Rev 39:1388–1405
13. Spiteri C, Moses JE (2010) Copper-catalyzed azide-alkyne cycloaddition: regioselective synthesis of 1, 4, 5-trisubstituted 1, 2, 3-triazoles. Angew Chem 122:31–33
14. Willis RJ, Marlow ID (1990) (*E*)-2-(4-Arylbut-1-en-3-yn-1-yl)chromones as Synthons for the Synthesis of Xanthone-1,2,3-triazole Dyads. Eur Pat Appl 400842
15. (a) Huisgen R (1963) 1,3-dipolar cycloadditions. past and future. Angew Chem Int Ed Engl 2:565–598; (b) Huisgen R (1963) Kinetics and mechanism of 1,3-dipolar cycloadditions. Angew Chem Int Ed Engl 2:633–645
16. (a) Tornøe CW, Christensen C, Meldal M (2002) Peptidotriazoles on solid phase: [1,2,3]-triazoles by regiospecific copper(I)-catalyzed 1,3-dipolar cycloadditions of terminal alkynes to azides. J Org Chem 67:3057–3064; (b) Rostovtsev VV, Green LG, Fokin VV, Sharpless KB (2002) A stepwise Huisgen cycloaddition process: copper(i)-catalyzed regioselective ligation of azides and terminal alkynes. Angew Chem Int Ed 41:2596–2599; (c) Zhang L, Chen X, Xue P, Sun HH, Williams ID, Sharpless KB, Fokin VV, Jia G (2005) Ruthenium-catalyzed cycloaddition of alkynes and organic azides. J Am Chem Soc 127:15998–15999; (d) Rasmussen LK, Boren BC, Fokin VV (2007) Ruthenium-catalyzed cycloaddition of aryl azides and alkynes. Org Lett 9:5337–5339

17. (a) Campos VR, Abreu PA, Castro HC, Rodrigues CR, Jordao AK, Ferreira VF, de Souza MC, Santos FDC, Moura LA, Domingos TS, Carvalho C, Sanchez EF, Fuly AL, Cunha AC (2009) Synthesis, biological, and theoretical evaluations of new 1,2,3-triazoles against the hemolytic profile of the Lachesis muta snake venom. Bioorg MedChem 17:7429–7434; (b) Zhou Y, Lecourt T, Micouin L (2010) Direct synthesis of 1,4-disubstituted-5-alumino-1,2,3-triazoles: copper-catalyzed cycloaddition of organic azides and mixed aluminum acetylides. Angew Chem 122: 2661–2664
18. (a) Ackermann L, Potukuchi HK, Landsberg D, Vicente R (2008) Copper-catalyzed "click" reaction/direct arylation sequence: modular syntheses of 1,2,3-triazoles. Org Lett 10:3081–3084; (b) Hein JE, Tripp JC, Krasnova LB, Sharpless KB, Fokin VV (2009) Copper(I)-catalyzed cycloaddition of organic azides and 1-iodoalkynes. Angew Chem 121:8162–8165
19. (a) Sletten EM, Bertozzi CR (2008) A hydrophilic azacyclooctyne for Cu-free click chemistry. Org Lett 10:3097–3099; (b) Jewett JC, Sletten EM, Bertozzi CR (2010) Rapid Cu-free click chemistry with readily synthesized biarylazacyclooctynones. J Am Chem Soc 132:3688–3690; (c) Arigela RK, Mandadapu AK, Sharma SK, Kumar B, Kundu B (2012) Cascade intermolecular Michael addition–intramolecular azide/internal alkyne 1,3-dipolar cycloaddition reaction in one pot. Org Lett 14:1804–1807; (d) Gold B, Dudley GB, Alabugin IV (2013) Moderating strain without sacrificing reactivity: design of fast and tunable noncatalyzed alkyne–azide cycloadditions via stereoelectronically controlled transition state stabilization. J Am Chem Soc 135:1558–1569
20. Tsai CW, Yang SC, Liu YM, Wu MJ (2009) Microwave-assisted cycloadditions of 2-alkynylbenzonitriles with sodium azide: selective synthesis of tetrazolo[5,1-a]pyridines and 4,5-disubstituted-2H-1,2,3-triazoles. Tetrahedron 65:8367–8372
21. Majireck MM, Weinreb SM (2006) A study of the scope and regioselectivity of the ruthenium-catalyzed [3 + 2]-cycloaddition of azides with internal alkynes. J Org Chem 71:8680–8683
22. (a) Huisgen R, Szeimies G, Mobius L (1966) Azide Chemistry—An Inorganic Perspective [3+2]-Cycloaddition Reactions of Metal Azides and Related Systems. Chem Ber 99:475–490; (b) Broeckx W, Overbergh N, Samyn C, Smets G, L'abbe G (1971) Cycloaddition reactions of azides with electron-poor olefins: isomerization and thermolysis of the resulting Δ2-triazolines. Tetrahedron 27:3527–3553
23. (a) Husinec S, Porter AEA, Roberts JS, Strachan CH (1984) Some Approaches to the Synthesis of Kainic Acid. J Chem Soc Perkin Trans 1:2517–2522; (b) Anderson GT, Henry JR, Weinreb SM (1991) High-pressure induced 1,3-dipolar cycloadditions of azides with electron-deficient olefins. J Org Chem 56: 6946–6948
24. (a) Amantini D, Fringuelli F, Piermatti O, Pizzo F, Zunino E, Vaccaro L (2005) Synthesis of 4-aryl-1H-1,2,3-triazoles through TBAF-catalyzed [3 + 2] cycloaddition of 2-aryl-1-nitroethenes with $TMSN_3$ under solvent-free conditions. J Org Chem 70:6526–6529; (b) Sengupta S, Duan H, Lu W, Petersen JL, Shi X (2008) One step cascade synthesis of 4,5-disubstituted-1,2,3-(NH)-triazoles. Org Lett 10:1493–1496; (c) WangY-C, Xie Yu-Y, Qu H.-En, Wang H.-S, Pan Y.-M, Huang Fu-P (2014) $Ce(OTf)_3$-catalyzed [3 + 2] cycloaddition of azides with nitroolefins: regioselective synthesis of 1,5-disubstituted 1,2,3-triazoles. J Org Chem 79: 4463–4469
25. Hansen SG, Jensen HH (2009) Microwave irradiation as an effective means of synthesizing unsubstituted N-linked 1,2,3-triazoles from vinyl acetate and azides. Synlett 3275–3278
26. Peng W, Zhu S (2003) Efficient synthesis of 5-fluoroalkylated 1H-1,2,3-triazoles and application of the bromodifluoromethylated triazole to the synthesis of novel bicyclic gem-difluorinated 1H-pyrano[3,4-d][1,2,3]-triazol-4-one compounds. Tetrahedron 59:4395–4404
27. (a) Kayet A, Pathak T (2013) 1,5-disubstituted 1,2,3-triazolylation at C1, C2, C3, C4, and C6 of pyranosides: a metal-free route to triazolylated monosaccharides and triazole-linked disaccharides. J Org Chem 78:9865–9875; (b) Sahu D, Dey S, Pathak T, Ganguly B (2014) Regioselectivity of vinyl sulfone based 1,3-dipolar cycloaddition reactions with sugar azides by computational and experimental studies. Org Lett 16:2100–2103

28. (a) Peng W, Zhu C (2003) Efficient synthesis of 5-fluoroalkylated 1*H*-1,2,3-triazoles and application of the bromodifluoromethylated triazole to the synthesis of novel bicyclic gem-difluorinated 1*H*-pyrano[3,4-d][1,2,3]-triazol-4-one compounds. Tetrahedron 59:4395–4404; (b) Roque DR, Neill JL, Antoon JW, Stevens EP (2005) Synthesis of 1,2,3-Triazoles by Cycloadditions of Azides with Enol Ethers. Synthesis 2497–2502
29. Quiclet-Sire B, Zard SZ (2005) The Synthesis of 1,2,3-Triazoles from Nitroalkenes—Revisited. Synthesis 19:3319–3326
30. Sengupta S, Duan H, Lu W, Petersen JL, Shi X (2008) One Step Cascade Synthesis of 4,5-Disubstituted-1,2,3-(*NH*)-Triazoles. Org Lett 10:1493–1494
31. Janreddy D, Kavala V, Kuo C-W, Chen W-C, Ramesh C, Kotipalli T, Kuo T-S, Chen M-L, He C-H, Yao C-F (2013) Copper(I)-Catalyzed Aerobic Oxidative Azide–Alkene Cycloaddition: An Efficient Synthesis of Substituted 1,2,3-Triazoles. Adv Synth Catal 355: 2918–2927
32. Xie Y-Y, Wang Y-C, Qu H-E, Tan X-C, Wang H-S, Pan Y-M (2014) Regioselective Synthesis of β-Aryl Enaminones and 1,4,5- Trisubstituted 1,2,3-Triazoles from Chalcones and Benzyl Azides. Adv Synth Catal 356:3347–3355
33. Zhang Y, Li X, Li J, Chen J, Meng X, Zhao M, Chen B (2012) CuO-Promoted Construction of *N*-2-Aryl-Substituted-1,2,3-Triazoles via Azide-Chalcone Oxidative Cycloaddition and Post-Triazole Arylation. Org Lett 14:26–29
34. Chen Y, Nie G, Zhang Q, Ma S, Li H, Hu Q (2015) Copper-Catalyzed [3+2] Cycloaddition/ Oxidation Reactions between Nitro-olefins and Organic Azides: Highly Regioselective Synthesis of NO2-Substituted 1,2,3-Triazoles. Org Lett 17:1118–1121
35. Rohilla S, Patel SS, Jain N (2016) Copper Acetate Catalyzed Regioselective Synthesis of Substituted 1,2,3-Triazoles: A Versatile Azide–Alkene Cycloaddition/Oxidation Approach. Eur J Org Chem 847–854
36. Yang B, Zou MC, Chen F, Fan KJ (2015) CuBr-Catalyzed Synthesis of 1,4,5-Trisubstituted 1,2,3-Triazoles Through Cycloaddition of Azides to Alkenes. Synth Commun 45:386–391
37. (a) Gangaprasad D, Raj JP, Kiranmye T, SagubarSadik S, Elangovan J (2015) A new paradigm of copper oxide nanoparticles catalyzed reactions: synthesis of 1,2,3-triazoles through oxidative azide-olefin cycloaddition. RSC Adv 5:63473–63477; (b) Gangaprasad D, Raj JP, Kiranmye T, Sasikala R, Karthikeyan K, Rani SK, Elangovan J (2016) A tunable route to oxidative and eliminative [3+2] cycloadditions of organic azides with nitroolefins: CuO nanoparticles catalyzed synthesis of 1,2,3-triazoles under solvent-free condition. Tetrahedron Lett 57:3105–3108
38. Alonso F, Moglie Y, Radivoy G, Yus M (2013) Alkenes as Azido Precursors for the One-Pot Synthesis of 1,2,3-Triazoles Catalyzed by Copper Nanoparticles on Activated Carbon. J Org Chem 78:5031–5037
39. Kamal A, Swapna P (2013) An improved iron-mediated synthesis of *N*-2-aryl substituted 1,2,3-triazoles. RSC Adv 3:7419–7426
40. Zhang J, Chang CWT (2009) Divergent Synthesis of Three Classes of Aryl *N*-Glycosides by Solvent Control. Org Chem 74:685–695
41. Marminon C, Gentili J, Barret R, Nebois P (2007) Synthesis of *N*-benzylated indole-, indazole- and benzotriazole-4,7-diones. Tetrahedron 63:735–739
42. (a) Ramachary DB, Ramakumar K, Narayana VV (2008) Amino Acid-Catalyzed Cascade [3+2]-Cycloaddition/Hydrolysis Reactions Based on the Push–Pull Dienamine Platform: Synthesis of Highly Functionalized *N*H-1,2,3-Triazoles. Chem Eur J 14:9143–9147; (b) Belkheira M, Abed DE, Pons JM, Bressy C (2011) Organocatalytic Synthesis of 1,2,3-Triazoles from Unactivated Ketones and Arylazides. Chem Eur J 17:12917–12921; (c) Danence LJT, Gao Y, Li M, Huang Y, Wang J (2011) Organocatalytic Enamide–Azide Cycloaddition Reactions: Regiospecific Synthesis of 1,4,5-Trisubstituted-1,2,3-Triazoles. Chem Eur J 17:3584–3587; (d) Wang L, Peng S, Danence LJT, Gao Y, Wang J (2012) Amine-Catalyzed [3+2] Huisgen Cycloaddition Strategy for the Efficient Assembly of Highly Substituted 1,2,3-Triazoles Chem Eur J 18:6088–6093; (e) Ramachary DB, Shashank AB (2013) Organocatalytic Triazole Formation, Followed by Oxidative Aromatization:

Regioselective Metal-Free Synthesis of Benzotriazoles Chem Eur J 19:13175–13181; (f) Yeung DKJ, Gao T, Huang J, Sun S, Guo H, Wang J (2013) Organocatalytic 1,3-dipolar cycloaddition reactions of ketones and azides with water as a solvent. Green Chem 15:2384–2388; (g) Li W, Jia Q, Du Z, Wang J (2013) Direct access to triazole-olefins through catalytic cycloaddition of azides to unsaturated aldehydes Chem Commun 49:10187–10189; (h) Li W, Du Z, Huang J, Jia Q, Zhang K, Wang J (2014) Direct access to 1,2,3-triazoles through organocatalytic 1,3-dipolar cycloaddition reaction of allyl ketones with azides.Green Chem 16:3003–3006; (i) Saraiva MT, Costa GP, Seus N, Schumacher RF, Perin G, Paixão MW, Luque R, Alves D (2015) Room-Temperature Organocatalytic Cycloaddition of Azides with β-Keto Sulfones: Toward Sulfonyl-1,2,3-triazoles. Org Lett 17:6206–6209; (j) Li W, Du Z, Zhang K, Wang J (2015) Organocatalytic 1,3-dipolar cycloaddition reaction of α,β-unsaturated ketones with azides through iminium catalysis. Green Chem 17:781–784

43. Wang T, Guo Z (2006) Copper in medicine: homeostasis, chelation therapy and antitumor drug design. Curr Med Chem 13:525–538
44. Yang W, Miao T, Li P, Wang L (2015) Regioselective synthesis of triazoles via base-promoted oxidative cycloaddition of chalcones with azides in aqueous solution. RSC Adv 5:95833–95839
45. (a) Singh N, Pandey SK, Tripathi RP (2010) Regioselective [3+2] cycloaddition of chalcones with a sugar azide: easy access to 1-(5-deoxy-D-xylofuranos-5-yl)-4,5-disubstituted-1*H*-1,2,3-triazoles Carbohydr Res 345:1641–1648; (b) Ajay A, Sharma S, Gupt MP, Bajpai V, Hamidullah KB, Kaushik MP, Konwar R, Ampapathi RS, Tripathi RP (2012) Diversity Oriented Synthesis of Pyran Based Polyfunctional Stereogenic Macrocyles and Their Conformational Studies Org Lett 14:4306–4309; (c) Ajay A, Prasad M, Devender GN, Tripathi RP (2012) Accessing a small library of pluripotent 1,4,5-trisubstituted 1*H*-1,2,3-triazoles via diversity-oriented synthesis Molecular Diversity 16:335–350
46. (a) Ramachary DB, Ramakumar K, Narayana VV (2008) Amino Acid-Catalyzed Cascade [3+2]-Cycloaddition/Hydrolysis Reactions Based on the Push–Pull Dienamine Platform: Synthesis of Highly Functionalized *N*H-1,2,3-Triazoles Chem Eur J 14:9143–9147; (b) Belkheira M, Abed DE, Pons JM, Bressy C (2011) Organocatalytic Synthesis of 1,2,3-Triazoles from Unactivated Ketones and Arylazides Chem Eur J 17:12917–12921
47. (a) Ramachary DB, Shashank AB, Karthik S (2014) An Organocatalytic Azide–Aldehyde [3+2] Cycloaddition: High-Yielding Regioselective Synthesis of 1,4-Disubstituted 1,2,3-Triazoles. Angew Chem Int Ed 53:10420–10424; (b) Ramachary DB, Shashank AB, Karthik S (2014) An Organocatalytic Azide–Aldehyde [3+2] Cycloaddition: High-Yielding Regioselective Synthesis of 1,4-Disubstituted 1,2,3-Triazoles. Angew Chem 126:10588–10592; (c) Shashank AB, Karthik S, Madhavachary R, Ramachary DB (2014) An Enolate-Mediated Organocatalytic Azide–Ketone [3+2]-Cycloaddition Reaction: Regioselective High-Yielding Synthesis of Fully Decorated 1,2,3-Triazoles. Chem Eur J 20:16877–16881; (d) Ramachary DB, Krishna PM, Gujral J, Reddy GS (2015) An Organocatalytic Regiospecific Synthesis of 1,5-Disubstituted 4-Thio-1,2,3-triazoles and 1,5-Disubstituted 1,2,3-Triazoles. Chem Eur J 21:16775–16780; (e) Zhou X, Xu X, Liu K, Gao H, Wang W, Li W (2016) Organocatalytic 1,3-Dipolar Cycloaddition Reaction of β-Keto Amides with Azides—Direct Access to 1,4,5-Trisubstituted 1,2,3-Triazole-4-carboxamides. Eur J Org Chem 1886–1890
48. Li W, Jia Q, Du Z, Wang J (2013) Direct access to triazole-olefins through catalytic cycloaddition of azides to unsaturated aldehydes. Chem Commun 49:10187–10189

Phytochemical and Antibacterial Analysis of *Terminalia chebula* and *Terminalia bellirica*

Rama Sharma, Saloni Raizada, Anjali Gautam and A. K. Bhatia

Abstract Chemistry of Natural product is an ancient science. Secondary products present in the plants are responsible for beneficial medicinal effects of plant materials. This results from the combinations of these secondary metabolites. Harad is a species of *Terminalia*. *Terminalia chebula* (Harad) is one of the imperative herbs used for digestive problems. Harad is a magical herb and its fruit is used to treat acidity, heat burn, heart disease, constipation, ulcers, piles, inflammation, dysentery and diarrhea. Harad is considered as a remarkable medicine for its healing properties. It is also used to remove toxins from body. It is good for lungs, bronchitis and sinus. *Terminalia bellirica*, the Behada tree has medicinal values and uses in Ayurvedicpathy to removes all the three doshas. Behada fruit's skin is beneficial in case of anemia, jaundice and white leprosy. The seeds Behada control vomiting, cures bronchitis. Baheda cure disorders of blood, along with voice disruption. Baheda is laxative, mild hot and destroys germs and eye diseases. Based on medicinal importance, an effort has been made to review the antibacterial properties and phytochemical analysis of *Terminalia chebula* (Harad), *Terminalia bellirica* (*Baheda*). *S. aureus and E. coli bacterial strains were used* for antibacterial activity. The antibacterial activity was determined by well diffusion method. The extract of *Terminalia bellirica* was more active against S. aureus than *Terminalia chebula*. Against E. coli both extracts showed almost same zone of inhibition.

Keywords Terminalia chebula (Harad) · Terminalia bellirica (Baheda)
Phytochemical · Antibacterial study

R. Sharma (✉) · S. Raizada · A. Gautam · A. K. Bhatia
Department of Biotechnology, G.L.A. University, Mathura, India
e-mail: dr.rama76@gmail.com

V. S. Parmar et al. (eds.), *Green Chemistry in Environmental Sustainability and Chemical Education*, https://doi.org/10.1007/978-981-10-8390-7_13

1 Introduction

From the ancient time these medicinal plants are being used for medical practices and now green-chemistry steppen to make a skillful and early tool for natural drug development. Natural products are important origin for biologically active drugs [1]. Extraction of biologically active compound from medicinal plant allows the demonstration of their physiological activity. It also makes possible pharmacology studied leading to synthesis of a more prospective drug with lower toxicity [2]

Harad is a species of *Terminalia*, native to southern Asia from India and Nepal east to southwestern China (Yunnan), and south to Sri Lanka, Malaysia and Vietnam. It is a deciduous tree growing to 30 m (98 ft.) tall. The fruit is drupe-like, 2–4.5 cm long and 1.2–2.5 cm broad, blackish, with five longitudinal ridges. Digestive problem scan be cured by this valuable herb. The fruit of this magical herb is used to treat acidity, convulsion, ulcers and diarrhea. For each of the problems the quantity and usage differs but Harad is an essential constituent in most of the concoction or powder that ayurvedic physician use for stomach related ailments. Haritaki is a Sanskrit word used for harad, which means arise from God's home. Besides digestive cure Harad is also believed as a valuable medicine for its mitigate properties. It is also important in removing toxins from body (Fig. 1).

Behada is most beneficial for eye diseases, hair, for soreness of throat, nose problems, purifies blood, cough, anemia, heart disease, jaundice, white Leprosy and kills germs. It is distinctive medicine used for stomach disorders. The seeds of Behada fruit are bitter, control vomiting, and cure bronchitis. Its seeds reduce pain and swelling, but excess quantity causes vomiting. Baheda is a large deciduous tree common on plains and lower hills in Southeast Asia. Oil content (40%) of Terminalia bellirica seeds, whose fatty-acid methyl ester fulfil all of the major biodiesel requirements in the USA, German and European Union (Fig. 2).

Fig. 1 Harad fruit

Fig. 2 Baheda fruit

2 Results

Based on medicinal importance, an effort has been made to review the antibacterial properties and phytochemical analysis of *Terminalia chebula* (Harad) *Terminalia bellirica (Baheda). S. aureus and E. coli bacterial strains were used* for antibacterial activity. The antibacterial activity was determined by well diffusion method. The extract of *Terminalia bellirica* was more active against S. aureus than *Terminalia chebula*. Against E. coli both extracts showed almost same zone of inhibition (Figs. 3, 4, 5 and 6; Tables 1, 2 and 3).

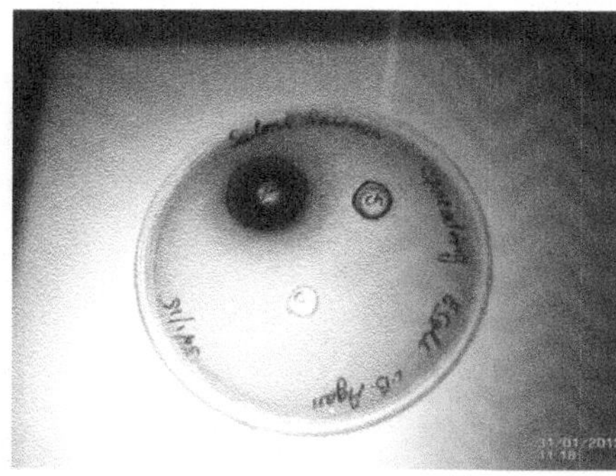

Fig. 3 Antibiogram of Harad against *E. coli*

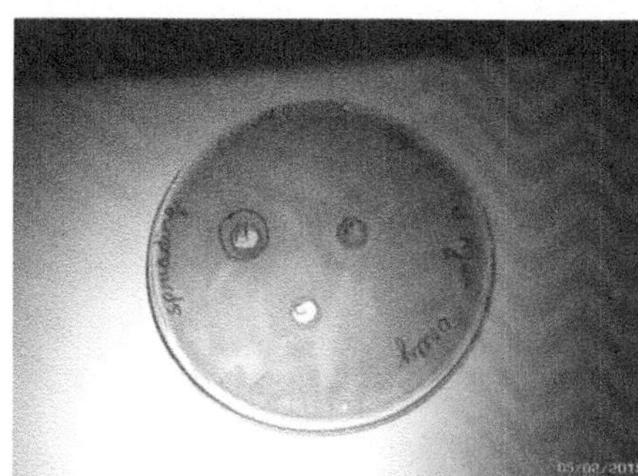

Fig. 4 Antibiogram of Baheda against *E. coli*

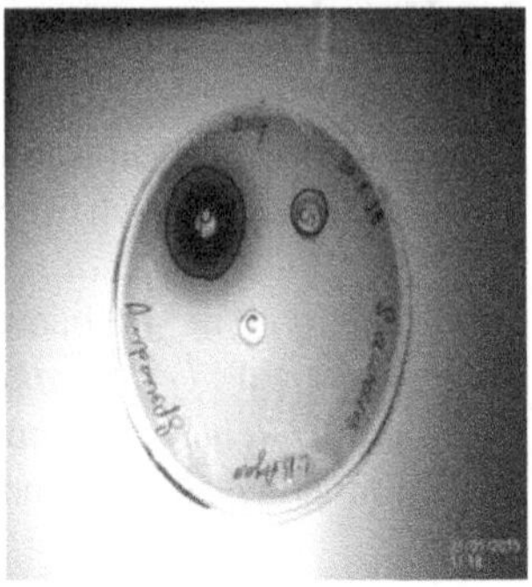

Fig. 5 Antibiogram of Harad against *S. aureus*

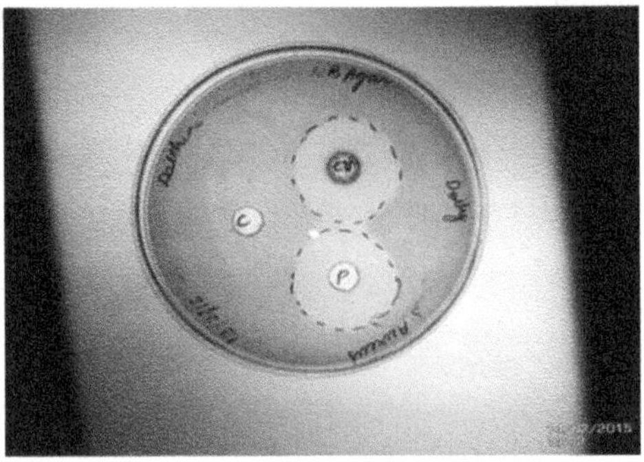

Fig. 6 Antibiogram of Baheda against *S. aureus*

Table 1 Zone of inhibition of plants against *E. coli*

Samples	Zone of inhibition (cm)
Harad	2.0
Baheda	2.2

Table 2 Zone of inhibition of plants against *S. aureus*

Samples	Zone of inhibition (cm)
Harad	1.3
Baheda	2.6

Table 3 Phytochemical test

Samples name	Harad	Baheda
Tests name		
Saponins	+ve	−ve
Tannins	+ve	+ve
Flavenoid	−ve	−ve
Terpenoid	+ve	+ve
Napthoquinine	−ve	−ve
Inulin	+ve	+ve
Glycosides	+ve	+ve
Carbohydrate	+ve	+ve
Alkaloids	+ve	+ve
Soluble phenols	+ve	+ve
Phlobotannins	+ve	+ve
Starch	−ve	−ve
Reducing sugar	+ve	+ve

3 Discussion

Both the extracts showed varying degrees of antimicrobial activity on the microorganisms tested. These plants could be a source of new antibiotic compounds. The results of these study showed that these medicinal plants generally used by standard medical professionals, were active against bacterial strains. Present study indicated variation in antimicrobial activities of selected samples. The antibacterial activity was determined by well diffusion method. The extract of *Terminalia bellirica* was more active against *S. aureus* than *Terminalia chebula.* Against *E. coli* both extracts showed almost same zone of inhibition. The present study has unveiled the importance of natural products to control antibiotic resistant bacteria which are being a danger to human health. This systematic study can fulfil as an important platform for the progress of economical, safe and potent medicines.

4 Materials and Methods

4.1 Collection of Samples

Harad and Baheda were taken to conduct the protocol. These were purchased from the market in mathura (Uttar Pradesh).

4.2 Preparation of Media

All media were prepared as per standard formulation given in Bacteriology Manual [3]. The dry ingredients were placed in a beaker, suspended in distilled water and then dissolved the medium completely. The prepared medium was dispended into flask and test tube, and finally sterilized by autoclaving at 121 °C for 30 min. About 15 mL was poured in Petri dishes aseptically. The plates were incubated at 37 °C for 24–48 h for sterility checking.

4.3 Collection of Bacteria

Procurement from Microbiology Lab of GLA University.

4.4 Preparation of Plant Extract

4.4.1 Collection of Plant Material

The barks were purchased from market. The selected plant materials were washed with clean water and allow to shade dried for about 2–3 weeks. The dried materials were crushed in an electric grinder to coarse powder.

4.4.2 Extraction of Plant Materials

Crude plant extract was prepared by Soxhlet extraction method as following [4].

Powdered material (40 g) was packed into a thimble and run in Soxhlet extractor. It was extracted with 200 mL methanol for about 48 h or till the solvent in the siphon tube of an extractor become colorless. After that extract was filtered with the help of filter paper and solvent evaporated from extract in Rotary evaporator to get the syrupy consistency. Anhydrous sodium sulphate was used to dried the residue and to remove trace of alcohol. This extract then kept in refrigerator at 4 °C and analyzed their physical and chemical property.

Antibiogram-well diffusion method was used to test antibacterial activity of plant extract [5].

Phytochemical Analysis: standard method was used for phytochemical analysis [6].

5 Conclusions

This study concluded that among Terminalia bellirica and Terminalia chebula, the extract of Terminalia bellirica is more active against S. aureus than Terminalia chebula while both extract show almost same zone of inhibition against E. coli.

Conflicts of Interest The authors declare no conflict of interest.

References

1. Lukmanul HF, Girija A, Boopathy R (2008) Antioxidant property selected ocimum species and their secondary metabolite content. J Med Plants Res 2(9):250–257
2. Manna A, Abalaka ME (2000) Spectrum J 7:119–125
3. AnejaK R (2003) Experiments in microbiology plant pathology and biotechnology, 4th edn. New Age International Publishers, New Delhi, pp 147–156
4. Okeke MI, Iroegbu CU, Eze EN, Okoli AS, Esimone CO (2001) Evaluation of extracts of the root of *Landolphia owerrience* for antibacterial activity. J Ethnopharmacol 78(119):127
5. Bauer AW, Kirby WMM, Sherris JC, Turck M (1966) Antibiotic susceptibility testing by a standardized single disk method. Am J Clin Pathol 36:493–496
6. Debela A (2002) Manual for phytochemical screening of medicinal plants. Ethiopian Health and Nutrition Research Institute, Addis Ababa, pp 35–47

Addressing the Environmental Feasibility of Electric Rickshaws

Ritu Khanna, Aleena Khan, Harshiel Chahal and Astha Goyal

Abstract In recent times electric rickshaw has been applauded as an environmentally sound and clean mode of transport. In contrast, the conventional fuel vehicles have proved to be extremely detrimental to the environment owing to their huge carbon emissions. This study is an attempt to quantify the impact of four modes of urban transit—e rickshaws, auto rickshaws, mini buses and buses, based on a primary survey of 220 e-rickshaw drivers from different areas of Delhi and methods of calculating carbon emissions adopted globally. The results conclude that the average kg of carbon emission is the lowest for the e-rickshaws as compared to the other fuel vehicles. However, the extent of contribution of e-rickshaws in greenhouse gas reduction is compromised due to carbon emissions from electricity generation required for charging the battery of the vehicle. Other challenges include disposal and recycling of lead batteries, energy losses and metal depletion which have environmental consequences of their own. As a result, the feasibility of solar batteries in e-rickshaws is explored. Finally, the paper suggests policy recommendations for improving the environmental profile of e-rickshaws. These include renewable sources for sustainable electricity generation, time of use electricity rates, a system of mother power stations to supply charged batteries to other local distribution points, and disposal and recycle systems of existing batteries.

Keywords E-rickshaw · Environmental effects · Carbon emission
Sustainable · Battery-disposal

R. Khanna (✉) · A. Khan · H. Chahal · A. Goyal
Department of Economics, Daulat Ram College, University of Delhi,
New Delhi 110007, Delhi, India
e-mail: ritukhanna01@gmail.com

A. Khan
e-mail: aleenarafikhan@gmail.com

H. Chahal
e-mail: chahalharshiel@gmail.com

A. Goyal
e-mail: astha96@yahoo.co.in

V. S. Parmar et al. (eds.), *Green Chemistry in Environmental Sustainability and Chemical Education*, https://doi.org/10.1007/978-981-10-8390-7_14

1 Introduction

The transportation sector is one of the major contributors to air pollution in urban areas. In India, passenger mobility mainly relies on rail and road and a large proportion of passengers use public transport for commuting. Public transport includes buses, metro, auto-rickshaws, cycle rickshaws etc. E-Rickshaws are the latest entry in this segment. The E-rickshaw was introduced to the urban public transportation system in 2010 when Municipal Corporation of Delhi launched them to solve the issue of last mile connectivity. It was introduced mainly in five areas—Greater Kailash II, Saket, IIT-SDA, DU-North Campus, and Chandni Chowk. An improved version of the cycle/manual rickshaws, they garnered huge popularity as accessible, cheap and comfortable modes of transport. As well as from the owner's perspective, these rickshaws are highly preferred due to their low cost of ownership and maintenance and higher returns. Most importantly, with pollution rates surging at an exceptional hight, e-rickshaw was seen as the best solution to the urban transportation related pollution problems. Thus, as per government figures, by April 2012, over 100,000 of these door-to-door connectivity providers had mushroomed in Delhi.

This paper is an attempt to study the environmental profile of e-rickshaws in comparison to three other conventional fuel vehicles—auto-rickshaws, buses and mini buses. The paper is divided into following sections. Section 2 describes the research methodology adopted in the study. Section 3 underlines the environmental costs of all four modes of transport. Section 4 deals with the demerits of lead battery used in e-rickshaw and explores the feasibility of lithium-ion solar battery as a more efficient alternative. Section 5 lays down the various policy recommendations to optimize the benefits of the e-rickshaw industry in Delhi.

For the purpose of this study, a primary research in the form of a survey of e-rickshaw drivers was conducted. Delhi was divided into four data sampling zones with a total of 220 e-rickshaws being sampled. A common questionnaire to evaluate their personal and vehicular profile was used and the results help in the afore-mentioned analysis.

2 Research Methodology

The most recent technique of measuring or quantifying the environmental degradation owing to vehicular pollution is the measurement of the kilograms of carbon emitted by a vehicle per passenger ($kgCO_2$ per km per passenger). This paper evaluates the carbon emission by an average e-rickshaw in Delhi by calculating the carbon emission due to the electricity used to charge the battery of the rickshaw.

To charge a battery of 650 W, 2.34 kJ of energy is required per hour (650 * 60 * 60 = 2,340,000 J/1000 = 2340–2.34 kJ/h). Further,

Table 1 $kgCO_2$ emissions by different modes of transport

Vehicle	$kgCO_2$ emission per passenger
Auto-rickshaw	0.01
Bus	0.004
Mini-bus	0.004

Source CO_2 emissions from passenger transport in India: 1950–51 to 2010–21, Sanjay Singh

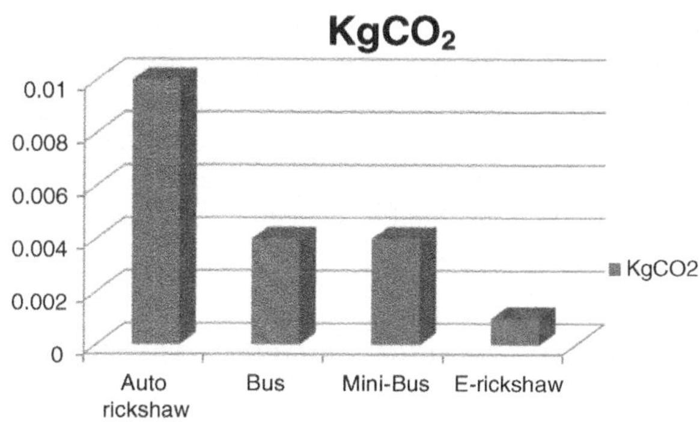

Fig. 1 Per passenger $kgCO_2$ emission by vehicles

$$1\ \text{kW} = 1000\ \text{J/s} = 3600\ \text{kJ/h}$$
$$2.34\ \text{kJ} = 2.34/3600 = 0.00065\ \text{kW/h}.$$

Now, 1 kW of energy produces 0.82 $kgCO_2$ (coolcalculator.org), hence per day carbon emission by e-rickshaw equals: 0.00065 * 0.82 * 12 = 0.064 $kgCO_2$.

According to the survey conducted, on an average an e-rickshaw carries around 59 passengers daily. Thus per passenger carbon emission by an e-rickshaw would be 0.064/59 = 0.0012 $kgCO_2$. The per passenger carbon emission by other conventional fuel vehicles is typical to calculate and hence in this paper referred from already calculated figures (Table 1).

Thus we conclude that auto-rickshaws (0.01 $kgCO_2$) emit the highest degree of $kgCO_2$ emission, followed by buses and mini buses (0.004 $kgCO_2$) and finally by e-rickshaws (0.0012 $kgCO_2$), emerging as the cleanest mode among all (Fig. 1).

3 Environmental Cost of the Four Modes

Conventional vehicles mostly operate on fossil fuels namely diesel, petrol, CHG and LPG emitting huge amounts of CO_2 into the atmosphere. The following section discusses the various environmental consequences of a few means of transport commonly used today in Delhi.

3.1 Environmental Consequences of CNG Auto-Rickshaws

Recent studies show that on an average only 14% of the auto-rickshaws have been converted into CNG operated vehicles. Extra lubricating oil is required to prevent the engine from piston seizure due to overheating. On an average around 200 ml of oil is used daily and this oil is released in the air after burning incompletely in the engine. The pollution caused by oil combustion is in addition to the emission due to CNG combustion and hence offsetting the advantages of using the CNG engine. Maintaining the CNG engine as compared to the petrol engine is more expensive and difficult these days. The head and piston are not adequately lubricated in the new engine and, therefore, an additional cooling requirement is needed. Furthermore, the new drive through a chain creates recurrent problems because the chain is not heavy-duty and requires regular adjustment. When the rings and chains wear out, oil consumption increases to 800 ml per day.

3.2 Environmental Consequences of Diesel Buses

The most commonly used engine today is the diesel engine. But diesel also greatly contributes to pollution worldwide, and this scenario is there to stay, with the large increases expected in vehicle count and vehicle miles travelled (VMT) leading to accelerated emissions globally. Delhi has an enormous number of buses running on its roads; on an average the ***kgCO_2 emission of Bus and Mini Bus is 0.004 and 0.004 kg per km per passenger respectively***. Besides contributing to global climate change, diesel emissions also lead to development of various diseases like cancer, cardiovascular and respiratory health effects. Vehicle idling as a result of long traffic jams emit noteworthy quantities of pollutants such as carbon dioxide, nitrogen oxides and volatile organic compounds (NOx and VOCs contribute to the formation of ozone smog), carbon monoxide, and particulate matter (PM leads to asthma, heart disease, lung damage, and possibly cancer). Wastage of fuel, excessive engine wear, and threat to many health risks are some of the other concerns with vehicle idling.

3.3 Environmental Consequences of E-Rickshaw

Operating solely on electricity, e-rickshaw is popular as one of the cleanest modes of transport today and is positively solving the problem of 'last minute connectivity'. It operates on a lead-acid battery of 650 W of power. The disposal of lead battery and the energy requirements for charging them make the ultimate cleanliness of this vehicle questionable.

However it is important to note that the higher value of per ws (0.01) than by buses (0.004) despite CNG passenger kgCO_2 emission by CNG auto rickshabeing introduced as a clean fuel can be deceptive. This happens because of the difference in the average number of passengers that the two vehicles carry. Hence, even if the CO_2 emissions of the diesel bus is more, it's per passenger $kgCO_2$ is lower than auto-rickshaws.

4 Lead Acid Batteries in E-Rickshaws

Approximately 140,000 MT of lead is used by India per annum, of which 95% is utilised by the battery industry. If such quantities of lead is not used with precaution and left unmonitored it can lead to a lot of pollution and hence severe damage to life.

5 Recycling Lead Battery

When a lead battery gets inoperative, it is important to make sure its proper collection and eco-friendly recycling because lead is a highly noxious metal and if left unchecked can be fatal for organic life. The process by which used lead is recycled for use in a new battery is known as smelting. However, many local smelters use the crude smelting processes that are not ecological. Thus, the need of the hour is to find a way of collecting old batteries and recycle them in an environment friendly process to avoid lead pollution. According to the survey conducted, almost 60% of the drivers exchange their batteries through registered dealers. It is a positive start although what proportion of batteries are being efficiently recycled is still ambiguous.

6 Lithium-Ion Solar Batteries

An emerging alternative for the lead acid battery is the lithium-ion solar batteries. These batteries are 1/3rd in weight and their life cycle time is 5000 times higher than the lead batteries. They are 100% efficient in charge and discharge even at low temperatures and the voltage is maintained throughout the discharge. Lithium-ion batteries are a breakthrough that if exploited to their full potential would change the face of Indian roads in the coming years.

7 Policy Recommendations

(a) **Efficient Recycling System**

Recycling of lead-acid batteries in an environmentally sound manner is important. If lead batteries are left unattended, they can contaminate the groundwater. Moreover, instead of producing fresh lead from ore, extracting it from old batteries is easier and requires much less energy. This paper suggests that an efficient system should be formalized where the drivers can deposit their old batteries in exchange of new ones and those used batteries can be transported to a place where it is recycled in an eco-friendly manner.

(b) **Mother Charging Stations**

There are various points that should be kept in mind while designing the recharge infrastructure. Firstly, the battery swap process should be as efficient and effective as possible and should take just the same amount of time as it takes to refuel petrol. Secondly, the economic goal is to design a system that maximizes the amount of energy to be produced by the energy source, while simultaneously minimizing the time to recover the invested money. Therefore, a centralized recharging station would be perfect. This plan suggests that the batteries would be charged at a central location, henceforth, leading to easy maintenance by trained staff, and confirming the permanency of the program. Thirdly, to reduce the extra stress on the grid a separate mechanism has to be considered.

A variety of options were pondered upon for transferring batteries to central location from the drop-off location: (1) use of the existing electrical grid (2) construction of a separate electrical distribution grid (3) transport of batteries by trucks. Using the existing electricity grid is not a viable option for the above mentioned problem of already overflowing load on these grids. On the other hand, the construction of a separate grid is relatively expensive. Moreover, Research of the petrol (gas), Compressed Natural Gas (CNG) and Liquefied Petroleum Gas (LPG) stations show the possibilities of supporting the infrastructure for e- rickshaws, and also combining solar technology in current and future stations. The charged batteries then can be sent to "daughter" stations. These are the already present gas stations or registered dealers of battery where until the drivers come for a replacement the batteries are stored. Finally, the discharged batteries are taken back to the mother charging stations and the charged ones are stored in the station to be installed in the rickshaw.

Source Journal of Asian Electric Vehicles, Volume 6, Number 2, December 2008.

While designing a mother power station setup the following points are to be ideated upon. (i) For a complete efficiency of the infrastructure design, the use of all-electric converted trucks is suggested. (ii) The daily demand of energy from the station has to be calculated to be able to optimize the size of the recharging station.

(c) **Time of use electricity rates**

One of the crucial drawbacks of the e-rickshaw is its huge demand for electricity that adds to the forever soaring demands in the cities today. One way to manage this can be by specifying particular timings to charge the battery of the e-rickshaw. The concept of time of use electricity rates where the per unit electricity cost varies according to the demand of electricity and charges off-peak, mid-peak and on-peak prices respectively can even out the distribution of electricity demand.

(d) Another important step would be the introduction of e-rickshaws using more clean energy such as wind and solar. India being a temperate zoned country has abundance of these renewable resources which are still hugely untapped. In times of increasing demand for electricity, such a step would be a breakthrough.
(e) Considering that Delhi has a huge number of auto-rickshaws on its roads, removing all of them would imply a huge economic turmoil, Thus introducing an electric version of these rickshaws will not lead to mass unemployment and will mitigate the pollution levels as well.

8 Conclusion

The paper concludes that with the increase in the demand for public transport in light of urbanization, e-rickshaw is one of the cleanest modes available. However, the acid-lead battery used in the rickshaws still poses serious threats to the environment if not handled properly. The need of the hour is to bring e-rickshaw as the foreman in the transport industry, both legally as well as in terms of infrastructure. Developing an efficient system to charge and dispose of batteries is of utmost importance to make the e-rickshaw industry environmentally most suited vehicle for door to door transit.

Acknowledgements This study has been conducted as part of the Innovation Project DRC 310, under the guidance of Mr. V. Packirisamy (Add. Gen. Manager, Power Finance Corporation Ltd.) and supported by a grant from the University of Delhi.

References

1. Lukic S, Mulhall P, Emadi A (2008) Energy autonomous solar/battery auto rickshaw. J Asian Electr Veh 6.2 (2008)
2. E-rickshaws: the road ahead. CPPR policy brief (2014) vol 1, Series II, pp 1–7 (Print)

3. Mohan D, Roy D (2015) Operating on three wheels auto-rickshaw drivers of Delhi. Econ Polit Wkly 38(3):177–180 (Print)
4. Costab Á, Sandra M, Patricia B (2014) Comparing the use of small sized electric vehicles with diesel vans on city logistics. Procedia Soc Behav Sci 3:1265–1274 (Print)
5. Lloyd AC, Cackette TA (2001) Diesel engines: environmental impact and control. J Air Waste Manag Assoc 51(6):809–847
6. Guidelines to Defra's GHG conversion factors: methodology paper for transport emission factors of the UK Department of Environment (Defra)
7. Zulkefa Rian M, Mizanur Rahman ANM (2014) Study on power consumption and social aspects of battery operated auto-rickshaw. In: International conference on mechanical, industries and energy engineering, 26–27 Dec 2014, Bangladesh
8. Shilu T, von Schirnding YE, Prapamentol T (2000) Environmental lead exposure: a public health problem of global dimensions. Bull World Health Organ 78(9) (Genebra Jan 2000)

Use of Cost Effective Kitchen Ingredients in Acid-Base Titrations: A Greener Approach

Ritu Payal, Arti Jain and Priti Malhotra

Abstract The volumetric analysis allows the quantitative chemical analysis of a solution of the unknown concentration (acid/base) by titrating it with the solution of accurately known concentration (base/acid). For this purpose, chemicals indicators such as phenolphthalein, methyl orange, methyl red, congo red etc. are used. These indicators are expensive and some of them have toxic effects on users and can also cause environmental pollution. Therefore, an attempt has been made to investigate the indicator activity of some of the fruits, vegetables and kitchen spices. The current paper recommends the use of natural indicators such as *Curcuma longa (turmeric), Beta vulgaris (beetroot), Citrus × limon (lemon), Citrus reticulata (orange), Citrus limetta (sweet lime), Solanum lycopersicum (tomato), etc.* as a strategy for environment remediation, to reduce toxicity and to decrease the respective cost of the synthetic indicators. Various acid-base titrations were carried to ensure a sharp color change with respect to the pH of the medium. The promising results were obtained using comparative studies against standard synthetic indicators. Consequently, the authors suggest that these indicators can be used as pH indicators in the college laboratories for undergraduate courses.

Keywords Kitchen ingredients · pH indicators · Cost effective
Acid-base indicators · Quantitative analysis

R. Payal
Department of Chemistry, Rajdhani College, University of Delhi,
New Delhi 110007, Delhi, India
e-mail: ritupayal.10@gmail.com

A. Jain · P. Malhotra (✉)
Department of Chemistry, Daulat Ram College, University of Delhi,
Delhi 110007, India
e-mail: pritimalhotra21@gmail.com

A. Jain
e-mail: jainarti21@gmail.com

V. S. Parmar et al. (eds.), *Green Chemistry in Environmental Sustainability and Chemical Education*, https://doi.org/10.1007/978-981-10-8390-7_15

1 Introduction

One of the major uses of indicators is in acid-base titrations in schools, colleges, research areas as well as possesses wide variety of applications in our everyday lives. Indicators are used to determine the end point by showing a characteristics color change with respect to a change in pH of the medium (Henderson Hasselbalch equation). In most of the acid-base titrations, synthetic indicators are a common choice and exhibit a different range of colors at different pH range [1].

Synthetic indicators are expensive, pollute the environment, have hazardous effect on the human body. For example, phenolphthalein is mostly used in acid-base experiments, however, numerous scientists proved that phenolphthalein is carcinogenic, causes gene alterations and deregulates the centrosome duplication cycle causing centrosome amplification [2, 3]. Therefore, an attempt is made to exploit indicator from natural sources (Fig. 1). These alternatives would be cost effective, easily and widely available, eco-friendly in nature as well as exhibit distinct color changes with variation in pH [4, 5].

Natural indicators are usually anthocyanins, betaines, chlorophylls, curcuminoids, flavonoids, flavones, which are sensitive to pH variation and can be used as indicators [6]. These natural sources possess antiseptic, anti-oxidative, anti-inflammatory, anti-bacterial, and many other biological properties that are used in the treatment of various diseases [7, 8]. The present work highlights the utilization of the natural ingredients as acid-base indicator in various acid base titrations.

Fig. 1 Different kitchen ingredients used in the present study

2 Results and Discussion

The results proved that indicators used in the present investigations show a characteristic color change under acidic or basic solutions (Table 1). The acidic solution changes color from dark red to yellow upon addition of base in case of turmeric powder, from colorless to yellow in case of citrus fruits, and from violet to yellow in case of beetroot. In addition, various acid-base titrations were carried out and compared with phenolphthalein. The titration values obtained using natural ingredients were in corroboration with an equivalence point obtained by standard indicator phenolphthalein.

The color change in basic medium is attributed to the negative charge generated by successive loss of hydrogen that generates a hybrid structure, which in turn is resonance stabilized [1]. These indicators have chromophores that are capable of undergoing a rearrangement depending on the pH, and results in a modification of electron transitions (π-π^*), and consequently, their molecular orbital (i.e. HOMO and LUMO).

3 Materials and Methods

Reagents: Analytical grade HCl (0.1M), CH_3COOH (0.1M), Oxalic acid (0.1M), NaOH (0.1M), phenolphthalein and various natural indicators such as turmeric powder, lemon, orange, sweet lime, beetroot, tomato, were used in the present work. Different types of Acid-base titrations i.e. strong acid versus strong base, weak acid against the strong base and mixture of strong and weak acid against strong base were carried out in the current study.

Preparation: Turmeric powder was used as such. Citrus fruits were squeezed and their juice was filtered and used. Beetroot was washed, peeled and then grinded in the mixture grinder and filtered off using minimal amount of water.

Method: All required reagents and volumetric solutions were prepared as per standard. The experimental work was carried out using same sample quantities and same set of glass wares for all types of titrations. The titrations were performed using 10 mL of titrant (20 mL in case of mixture of acids) with three to four drops of natural indicator (a pinch in case of turmeric). A set of four experiments each for all types of acid-base titrations were carried out.

Apparatus: Volumetric flasks, conical flasks, burettes, pipettes, spatulas, test tubes, droppers, funnel, clamp stand, pH paper, measuring cylinders etc.

Table 1 Color change in acid-base titrations using different indicators

Titration of base versus	Indicators used						
	Phenol-phthalein	Turmeric Powder	Lemon	Orange	Sweet lime	Beet root	Tomato
	Colorless to pink[b]	Yellow to red[b]	Colorless to yellow[b]			Violet to yellow[b]	Colorless to wine red[b]
HCl[a]	9.7	9.9	9.8	10.0	9.9	9.6	10.3
CH_3COOH[a]	12.6	10.2	10.0	10.3	10.5	11.3	12.0
Oxalic acid[a]	10.9	11.0	10.9	11.2	11.2	11.3	11.5
HCl + CH_3COOH[a]	22.1	22.4	21.9	22.1	22.0	22.4	22.6
HCl + Oxalic acid[a]	20.4	20.4	20.6	20.6	20.3	20.2	20.6
Oxalic acid + CH_3COOH[a]	23.2	22.8	23.2	23.0	22.9	23.3	23.5

[a]Volume of acid used = 10 mL (20 mL in case of mixture of acids)
[b]Equivalence point (two or three equivalent values were considered)

4 Conclusions

From the result it is proved that the synthetic indicators could be replaced by the natural pigments in different acid-base titrations because they are cost effective, widely available, simple to prepare, and non-hazardous. These indicators gave sharp color change at the equivalence point and also, equivalence point obtained using natural indicators correlate well with the synthetic indicators. The authors proposed that these natural indicators can be used as a substitution to synthetic indicator by adopting the principles of green chemistry.

Acknowledgements The support from Rajdhani College and Daulat Ram College for providing the laboratory facility to carry out the experimental work is highly appreciated.

Conflicts of Interest Declare conflicts of interest or state "The authors declare no conflict of interest."

References

1. Pradeep DJ, Dave K (2013) A novel, inexpensive and less hazardous acid-base indicator. J Lab Chem Edu 1(2):34–38
2. Ferreira MA, Alveus AC, Costa MAC, Paul MI (1977) Naphthoquinone dimers and trimers from *Euclea natalensis*. Phytochemistry 16(1):117–120
3. Heard PL, Rubitski EE, Spellman RA, Schuler MJ (2013) Phenolphthalein induces centrosome amplification and tubulin depolymerization in vitro. Environ Mol Mutagen 54:308–316
4. Eze SO, Ogbuefi RA (2014) Analytical potentials of dye extracts from *Aspilia africana* (oramejula) flowers. Asian J Nat Appl Sci 3(1):54–60
5. Macuvele DL, Sithole GZ, Cesca K, Macuvele SL, Matsinhe JV (2016) Aqueous extracts of mozambican plants as alternative and environmentally safe acid-base indicators. Environ Sci Pollut Res Int 23(12):11639–11644
6. Ikan R (1991) Natural products: a laboratory guide. San Diego, CA
7. Vasudeva N, Sharma SK (2008) Post-coital antifertility activity of *Hibiscus rosa-sinensis* Linn roots. Evid Based Complement Alternat Med 5(1):91–94
8. Weigenand O, Hussein AA, Lall N, Meyer JJM (2004) Antibacterial activity of naphthoquinones and triterpenoids from Eucleanatalensis root bark. J Nat Prod 67(11):1936–1938

Recent Advances in Triazolyl Nucleosides

Smriti Srivastava, Vipin K. Maikhuri, Divya Mathur and Ashok K. Prasad

Abstract Nucleosides and its modified analogues are important components and possess key roles in various biological processes. Copper catalyzed 1,3-dipolar cycloadditions have been emerged as an excellent tool to join azide and alkyne moieties to form triazolyl compounds. Presence of triazole group enhances the therapeutic potential as well as fluorescence properties of the modified nucleosides. Owing to the designing and development of the new triazolyl nucleosides for their various biological and photophysical application at faster pace, there is a need to have knowledge of the recent developments in this area. The review herein highlights the various modifications, therapeutic importance and other applications of recently reported triazolyl-nucleosides.

1 Introduction

Nucleic acids are the vital components of all living organisms. Monomeric units of nucleic acids, which contain sugar and nucleobases are called nucleosides. In the recent past, a surplus of modified nucleosides with varied modification have been synthesized demonstrating numerous pharmacological properties such as antibacterial, antiviral, anticancer etc. [1]. Various modified nucleosides and oligonucleotides involving them have been synthesized over the years, for their use in drug

S. Srivastava · V. K. Maikhuri · A. K. Prasad (✉)
Department of Chemistry, University of Delhi, New Delhi 110007, Delhi, India
e-mail: ashokenzyme@gmail.com

S. Srivastava
e-mail: smritichem10@gmail.com

V. K. Maikhuri
e-mail: vipinkrmaikhuri@gmail.com

D. Mathur
Department of Chemistry, Daulat Ram College, University of Delhi,
Delhi 110007, India
e-mail: dmchem05@gmal.com

V. S. Parmar et al. (eds.), *Green Chemistry in Environmental Sustainability and Chemical Education*, https://doi.org/10.1007/978-981-10-8390-7_16

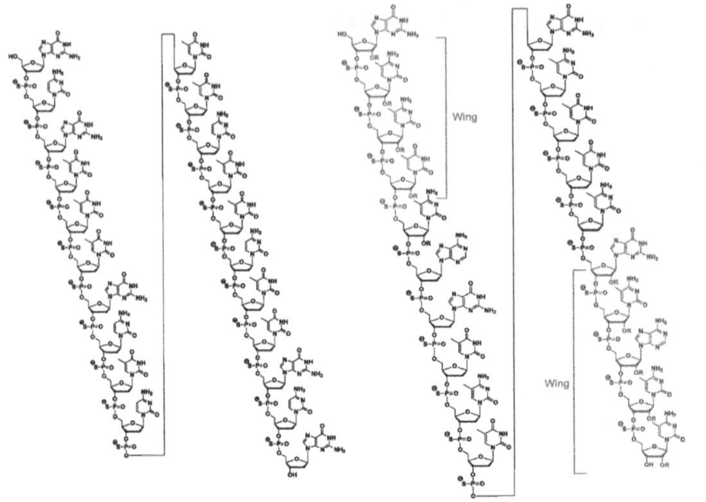

5'-GCGTTTGCTCTTCTTCTTGCG-3'
Vitravene (**1a**)

$5'\text{-}G_m{}^{Me}C_m{}^{Me}C_m{}^{Me}U_m{}^{Me}C_mAGT^{Me}CTG^{Me}CTT^{Me}CG_m{}^{Me}C_mA_m{}^{Me}C_m{}^{Me}C_m\text{-}3'$
Kynamro (**1b**)
(R = MOE; 2'-OMOE)

Fig. 1 Structure of Vitravene and Kynamro: antisense drugs approved by FDA

development. These modified nucleosides are important as oligonucleotides based on natural DNA/RNA possess few shortcomings, such as lesser nuclease resistance, poor binding affinity etc. perturbing their use in antisense therapeutics as such [2]. Modifications in the backbone [3], base [4] and sugar [5] functionalities of oligonucleotides have been carried out to overcome these limitations, resulting in significant progress towards establishing oligonucleotides as viable therapeutic agents. Vitravene [6], (**1a**) was the first antisense oligonucleotide having backbone modification approved for the treatment of retinitis, Kynamro [7] (**1b**) was the second antisense agent used for the cure of hypercholesterolemia which contained backbone, base and sugar modification (Fig. 1).

1.1 Copper Catalyzed Azide-Alkyne Cycloaddition (CuAAC)

The azide-alkyne cycloaddition reaction between an azide and a internal or terminal alkyne to give 1,2,3-triazoles was first introduced by Huisgen [8] but unfortunately this reaction requires elevated temperatures and produces mixture of regioisomers. Later on, in 2001, Meldal [9] and Sharpless [10] independently realized the importance of major improvement in rate and regioselectivity of the Huisgen cycloaddition reaction of azides and alkynes by using Cu(I) catalysis. 'Click reactions' can be defined as reactions which requires simple reaction conditions, high yielding, wide in scope, creates only inoffensive by-products and can be performed in the solvents which are easily removable or benign and give physiologically stable products. These reactions can be best defined as "all searches must be restricted to molecules that are easy to make", and so is the click reaction. Various research groups have widely used click chemistry as an important synthetic methodology for the production of pharmacologically active molecules.

1.2 Some of the Features of Click Reactions

- It is quantitative, robust, general, insensitive, wide in scope, easy to perform, uses only readily available reagents, easy workup and purification, uses eco-friendly solvents, used in orthogonal and biomolecular ligation reactions and is also used for in vivo tagging.
- It is one of the powerful reactions which form carbon-heteroatom-carbon bonds.
- The triazole ring formed is found to be chemically inert against various reaction conditions such as hydrolysis, oxidation and reduction. It possesses an intermediate polarity and has dipole moment of approx. 5D.
- These reactions are strongly exothermic either because of reactants which are strongly energetic or products having high stability.
- These reactions are generally fusion processes or condensation processes (thus leaving either no by-products or water as a by-product).

1.3 Role of Click Reaction in Nucleoside Chemistry

Modified nucleosides having triazole moeity have been synthesized by several research groups and were found to possess a diverse range of biological properties such as antibacterial [11], antifungal [12], anticancer [13] and antiviral [14] activities. Inspite of pharmacological properties, several triazolyl nucleosides also have photophysical and other applications. Triazole moieties can be formed at any position of the sugar ring as well as on the nucleobases.

2 Significance of Triazolyl Nucleosides

2.1 Pharmacological Activities of Triazolyl Nucleosides

Variously substituted 1,2,3-triazoles constitute one of the common moiety in several biologically active compounds and this had led to the development of numerous synthetic methodologies for their incorporation in more complex structures. Owing to the upsurge in the modified nucleosides for pharmacological applications, we documented here recent developments in the field of triazolyl nucleosides.

2.2 Anticancer Activity

Cancer is characterized by uncontrolled growth of cells, which have potential to infect various parts of the body. It is a serious life threatening problem worldwide. However, several novel and effective drugs for cancer treatment are present; there are still clear limitations present for its therapy. So, new drugs for cancer treatment are in great demands.

Alaoui et al. [15], synthesized a series of sulfonamide-4-substituted-1,2,3-triazolyl nucleosides and evaluated them against two tumor cell lines MDA-MB-231 and RCC4. Among all the synthesized compounds, **2** (Fig. 2) was found to be most potent compound with an IC_{50} of 13–15 μM, and at a dose of 50 μM, it completely eradicates MDA-MB-231and RCC4. Western blot analysis demonstrates that its mode of action involves apoptosis as well as autophagy of cell. Amdouni et al. [16], explored click/oxidative coupling and click electrophilic addition to synthesize a series of nucleoside analogues bearing a 1,4,5-trisubstituted-1,2,3-triazole aglycone (Fig. 3). These fully substituted 1,2,3-triazolyl-glycosides exhibit potent effect against various hematological malignancies like chronic myeloid leukemia (CML) and myelodysplastic syndrome (MDS) either resistance or sensitive to their reference therapy. One of the lead compound **3** was found to induce tumor retrogradation in nude mice xenotransplanted with azacitidine-resistant MDS cell lines without any significant toxic effects (Fig. 2). They also established that the antileukemic activity shown by the lead compound was due to its ability to induce both autophagy and caspase activation. Bodnár et al. [17], synthesized 2′-deoxynucleoside conjugates of 13α-esterone. All the synthesized compounds were evaluated against a group of human adherent cell lines (MCF-7, HeLa and A2780) for their in vitro antiproliferative activities. The protected cytidine 13α esterone conjugate **4** showed best antiproliferative activity (IC_{50} 9 μM). Giofrè et al. [18], reported the synthesis of 5-(1*H*-1,2,3-triazol-4-yl) isoxazolidines, a novel series of *C*-nucleosides mimicking the pyrimidine nucleobases due to the presence of isoxazolidine system which are linked via 1,2,3-triazole. All the synthesized compounds were evaluated for their antiproliferative activity and for compounds **5** and **6** the growth inhibitory effect at a 100 μM concentration after 72 h

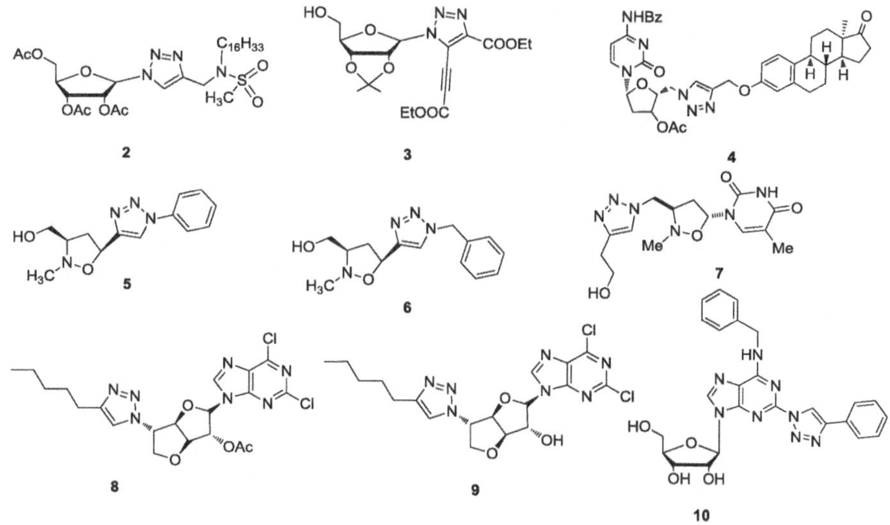

Fig. 2 Triazolyl nucleosides having anticancer activity

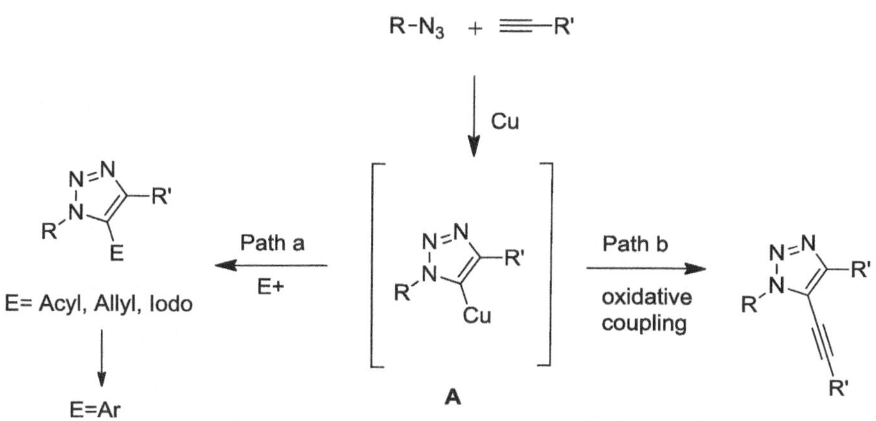

Fig. 3 Synthetic route for 1,2,3-trisubstituted triazolyl nucleosides

of incubation reaches up to 50% in HepG2 and HT-29 cells and increases up to 56% in the SH-SY5Y cell line. Romeo et al. [19], synthesized a series of 2′-oxa-3′-aza-4′-a-carbanucleoside, with a triazole linker at C-5′-position. Antitumor activity evaluation of these compounds against a panel of cell lines showed that most of them inhibit the proliferation of cells of human epidermoid carcinoma larynx cells (HEp-2), African green monkey kidney cells (Vero and BS-C-1), human foreskin fibroblast cells (HFF) and Madin–Darby canine kidney (MDCK) by 50% (CC_{50}) at concentration in the range of 5.0–40 μM. The lead compound **7** showed a high level of inhibitory activity with CC_{50} values of 5 μM for all the tested cell lines. A series of 2,6-dichloropurine bicyclonucleosides having a triazolyl carbohydrate moiety have been synthesized by Zang et al. [20] and were evaluated for their antitumor activity (**8** and **9**). These compounds showed significant inhibitory activity against PC-3, EC109, MGC-803 and HGC-7 cell lines. Lakshman et al. [21], synthesized *C*-2-1,2,3-triazolyl nucleoside derivatives as **10**, the reaction conditions for click reaction were optimized and CuCl in *t*-BuOH/H_2O (1:1) was found to be the optimal reaction condition. The desilylated products were used for the biological assays. A good anti-proliferative activity was showed by the *C*-2 triazolyl adenosine analogue.

2.3 Antiviral Activity

Viral diseases are extremely widespread infection caused by several types of viruses, a kind of microorganism. The diagnosis of these diseases has been done mostly but they are difficult to cure. Around the world, several human beings are suffering from the diseases caused by DNA and RNA viruses. Modified nucleosides can act as building blocks for DNA and RNA due to their similarity to the natural nucleosides. Drugs based on modified nucleosides therefore may play key roles to fight against viral infections. Ouahrouch et al. [22], described the I_2 catalyzed kabachnik-field reaction for the synthesis of 1,2,3-triazolylbenzyl aminophosphonates ribonucleosides. This reaction was followed by Cu (I) catalyzed click reaction under microwave irradiation. The activity of these synthesized compounds were evaluated against a broad range of DNA and RNA viruses, few of them showing moderate inhibitory activity against respiratory syncytial virus (RSV) (compound **11**) and Coxsackie virus B4 (compound **12**). Rashad et al. [23], synthesized some acyclic *C*-Nucleosides by treating 2-(5-phenyl-3H-1,2,4-triazol-3-ylsulfanyl)acetohydrazide with some aldoses. These compounds were tested for their antiviral activity and were found to be active against Hepatitis B virus. One of the lead compound **13** showed significant antiviral activity in comparison to the anti-influenza drug Zenamivir. Although, all of the tested compounds demonstrate no antiviral activity against herpetic viral type-1 and type-2 compared to the reference antiviral drug Acyclovir. 4′-azidothymidine (ADRT) were subjected to azide-alkyne cycloaddition reaction under Cu catalysis to give 1,2,3-triazole derivatives (**14**) by Verneker et al. [24]. Antiviral screening of these analogues were done, which showed these molecules to be moderately active against HIV-1 and or influenza A-virus and non-active against west nile virus (WNV) or hepatitis C-virus (HCV). Thus, they concluded that 4′-azido group might be important for the antiviral activity of these compounds and thus their chemical modification decreases the antiviral activity. Jamal et al. [25], synthesized a novel series of 5-(1,2,3-triazolyl)-2′-deoxyuridine using microwave assisted click reaction under Cu (I) catalysis. These compounds were screened for antiviral activities against human rhinvirus (HRV), HCV and Human Immunodeficiency Virus (HIV) and showed only marginal antiviral activities (**15**). Elayadi et al. [26], synthesized 1,4-disubstituted-1,2,3-triazole nucleosides by using montmorillonite K10 impregnated with potassium iodide and copper dichloride ($CuCl_2$/KI/k10) as a catalyst for the dipolar cycloaddition reaction in the solvent mixture of dioxane: MeOH (2:1). Few of the synthesised compounds were evaluated for their anti-viral activity in vitro. Compound **16** showed moderate inhibition against influenza virus A (H3N2). Zhou et al. [27]. have reported thymidine analogues containing 1,2,3-triazole **17** and evaluated their antiviral activities. Montague et al. [28] have synthesized *C*-5-(1,4- and 1,5-disubstituted-1,2,3-triazolo)-nucleoside derivatives **18** as antiviral agents. Vernekar et al. [29]. have synthesized 4′-(1,2,3-triazol-1-yl) thymidines **19** and determined their antiviral activity. Olomolo et al. [30], Synthesized coumarin triazolyl-conjugated nucleosides **20** and evaluated their potential against dual action HIV I protease and non-nucleoside reverse trancriptase. Hwu et al. [31] have synthesized coumarin-purine ribofuranoside conjugates **21** and evaluated them against HCV (Fig. 4).

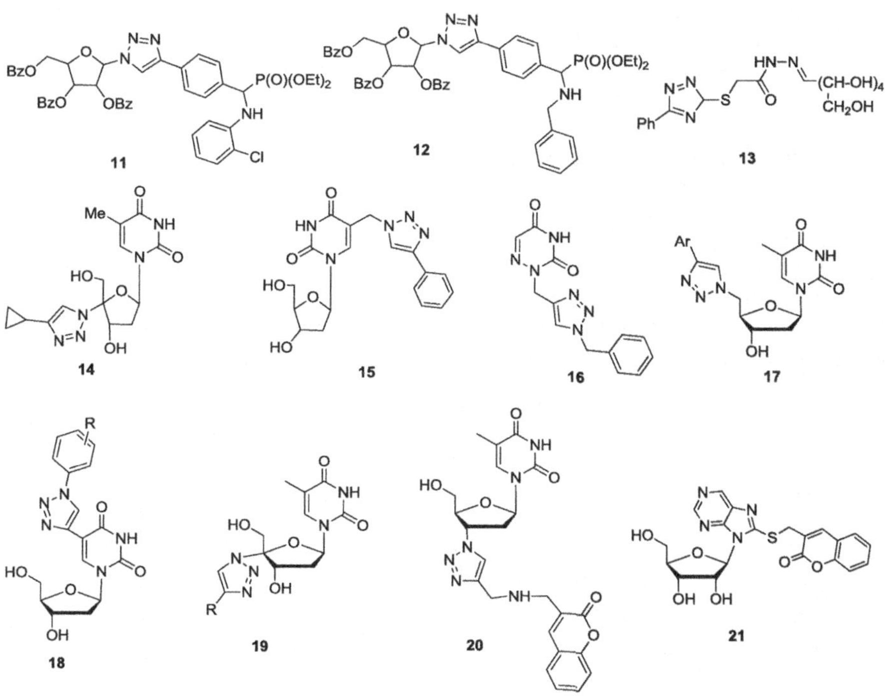

Fig. 4 Triazolyl nucleosides having antiviral activity

2.4 Enzyme Inhibitor Activity

Enzyme inhibitors are molecules that interact with enzymes in such a way to prevent it from working in the normal manner. Since blocking the activity of enzymes can kill a pathogen or correct the metabolic imbalances, enzyme inhibitors can be treated as drugs. The compounds synthesized by Bodnar et al. [17] for their anticancer activity evaluation were further evaluated for their inhibitory activity against human 17β-hydroxysteroid dehydrogenase 1 (17β-HSD1) via in vitro radiosubstrate incubation. The non-protected thymidine 13α estrone conjugate **22** showed best enzyme inhibition (IC_{50} = 19 μM). Xavier et al. [32], synthesized (triazolyl)methyl amide-linked disaccharide nucleoside. The non-protected or partly protected carbohydrate derivatives were taken as precursors. The carbohydrate moiety includes ribose, glucuronolactone, glucuronic acid and glycopyranosides. The reaction sequence includes stereoselective *N*-glycosylation, regioselective Mitsunobu coupling and followed by click chemistry. Few triazole-containing glyco-derivatives were found to have moderate selective acetylcholinesterase inhibitory activities. Also, the tested compounds were found to possess no toxicity to normal cells. Mahony et al. [33]. have synthesized 2′-([1,2,3]-triazol-1-yl)-2′-deoxyadenosine derivatives **23** as *t*-RNA synthetase inhibitor (Fig. 5).

2.5 Anti-leishmanial Activity

Leishmania is caused by the protozoan parasites of genus *Leishmania* having more than 20 different species. Presently, more than 80 countries around the world are infected by the parasite, leading to a large number of deaths every year. Although, this disease is a big threat to human health, still there is no effective treatment against leishmaniasis. So, there is a need for the development of new chemical

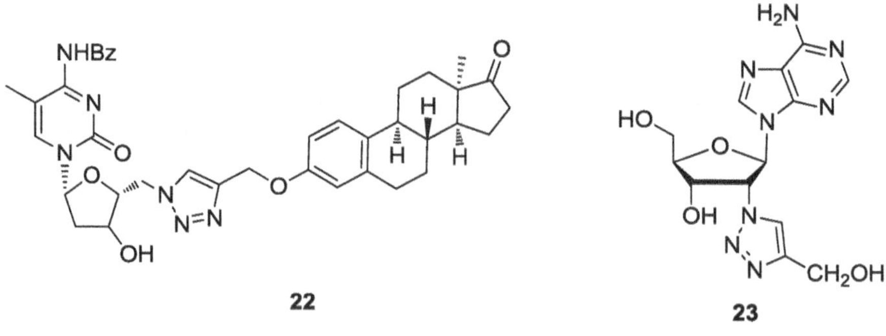

Fig. 5 Triazolyl nucleosides as enzyme inhibitors

Fig. 6 Triazolyl nucleoside having anti-leishmanial activity

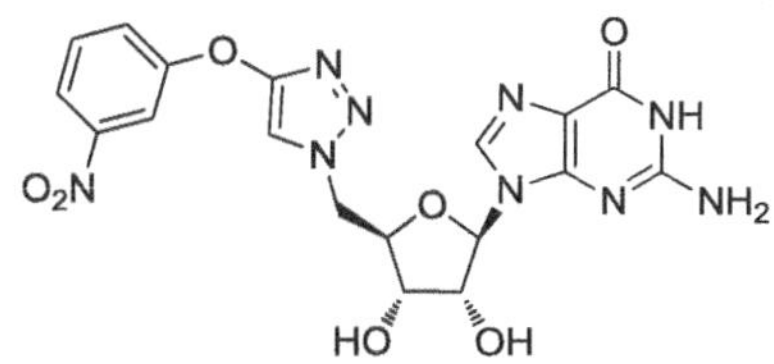

24

entities for the treatment of this disease. A series of 5′-aryl-5′-deoxy guanosine analogs were synthesized by Daligaux et al. [34]. by clicking together 5′-azido-5′-deoxy guanosine with different benzyloxy alkynes. The synthesized compounds were evaluated for in vitro activity against LV9 reference strain of axenic anastigotes and intermacrophase anastigotes of *Leishmania donovani.* Among all the screened compounds, 4-(3-nitrobenzyl)-1,2,3-triazole-5'-substituted guanosine analog (**24**) was found to be most active compound (Fig. 6).

2.6 Anti-bacterial Activity

Bacterial disease includes any kind of sickness caused by bacteria. Harmful bacteria causing bacterial infections are called pathogenic bacteria. Antibiotics are taken as medication for bacterial infections. However due to emergence of resistance, drug development for multidrug resistant bacteria has become important area of research. Jamal et al. [25], evaluated the antibacterial activity of compounds like 5-(1,2,3-triazolyl)-2′-deoxyuridine (**25**), against gram positive and gram negative bacteria. The screened compounds were found to have moderate anti-bacterial activity. Reddy et al. [35] synthesized tetrahydrofuranyl 1,2,3-triazolyl *C*-nucleoside analogues using click reaction between tetrahydrofuranyl alkynes with various sugar derived tetrahydrofuranyl azides and an admantyl azide. The synthesized *C*-nucleosides were evaluated for their anti-bacterial activity. The admantane containing 1,2,3-triazole *C*-nucleoside (**26**) was found to have MIC of 1.56 μg/mL against *S. aureus* and MIC of 0.78 μg/mL against *K. pneumonia.* The activity of compound **26** against *S. aureus* was about two fold higher than standard drug gentamycin and three fold higher than ampicillin. Also, in case of its activity against *K. pneumonia,* it was found to be equivalent to the activity of gentamycin. Another analogue containing adamantane (**27**) was found to have MIC of 0.78 μg/mL against *S. aureus* and MIC of 0.78 μg/mL against *K. pneumonia.* Its activity against *S. aureus* was found to be three fold higher than gentamycin, equivalent to ciprofloxacin and four fold higher than ampicillin. Also, its activity against *K. pneumonia* was equivalent to gentamycin (Fig. 7).

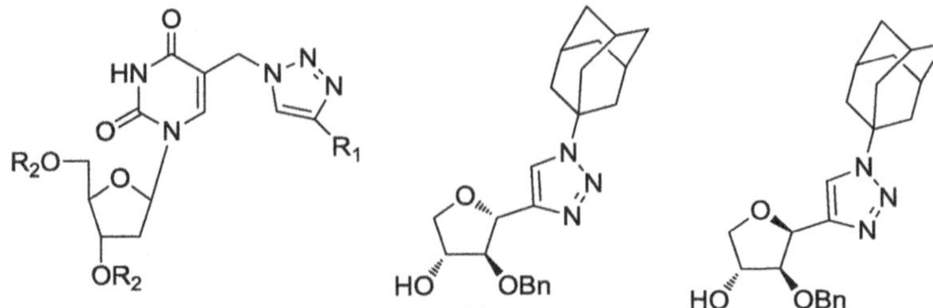

25 26 27

Fig. 7 Triazolyl nucleosides having anti-bacterial activity

2.7 *Anti-tubercular Activity*

Tuberculosis is an air borne disease caused by *Mycobacterium tuberculosis* that mainly affects the lungs and remains one of the most deadly infectious diseases for humans. World Health Organization (WHO) identified Tuberculosis as one of the three priority diseases for drug development research program [36]. The current TB therapies are not very effective and also have adverse effects. The emergence of TB strains having drug resistance aggravates the situation [37]. In this regards, several triazolyl nucleoside based antitubercular compounds have been designed by chemists.

Poecke et al. [38] reported *Mycobacterium tuberculosis* thymidine monophosphate kinase (TMPKmt) inhibitory activities of a series of newer analogues of 3′- and 5′-modified thymidine having both α and β-derivatives(**28** and **29**). It was reported by the authors that analogues with thymine base trans to the aromatic substituents showed the best inhibition of TMPKmt (Fig. 8).

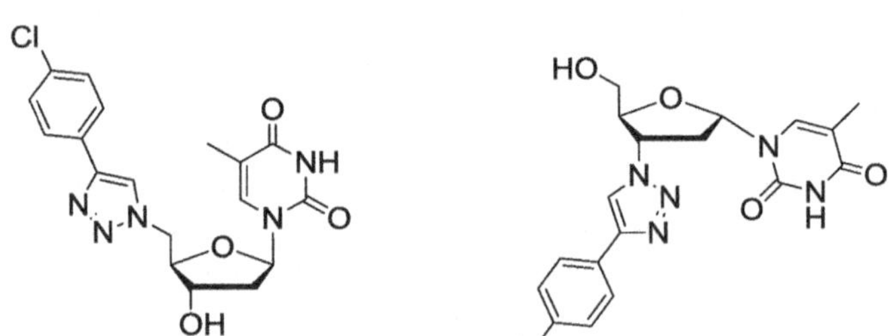

28 29

Fig. 8 Triazolyl nucleosides having anti-tubercular activity

2.8 Photophysical Properties

Modified nucleoside analogs showing fluorescence properties have been very helpful in the nucleic acid structure and function studies. These nucleoside analogs were incorporated into oligonucleotides, causing photophysical variations at the nucleotide level [39]. So, several research groups are interested nowadays to synthesize fluorescent triazolyl nucleoside and study their photophysical properties. Passays et al. [40], taking into account their previous report that some Ir (III) complexes were able to photoreact with DNA through photoinduced electron transfer, reported the synthesis of Ir (III) nucleosides using click reaction and their characterization. Natural H-bonded bases were replaced by metal-coordinated liganosides. They demonstrated that the affixing of Ir (III) complexes onto nucleoside through triazole linker was not altering the properties of these complexes and thus provide a new path for the study of DNA-charge transfer processes (**30**, Fig. 9). A spectroscopic and electrochemical investigation was done on the triazolyl Ir (III) nucleoside conjugates.

Bag et al. [41], designed the acceptor/donor chromophores with unnatural triazolyl nucleosides taking in consideration that a pair of such acceptor/donor nucleoside might be engaged in π-π stacking and photophysical interaction leading to the DNA duplex stabilization. They studied the photophysical properties of these acceptor/donor unnatural triazolyl nucleosides and found interesting salvatochromic fluorescence for two of the nucleosides. Also, they studied the photophysical interactions of a pair of α/β nucleosides, as well as between two acceptor-donor β-nucleosides. One of these triazolyl nucleosides (**31**) was further exploited for interaction studies with BSA. They determined that electrostatic and hydrophobic interaction played an important role for sensing of BSA by modified nucleoside **31** (Fig. 10).

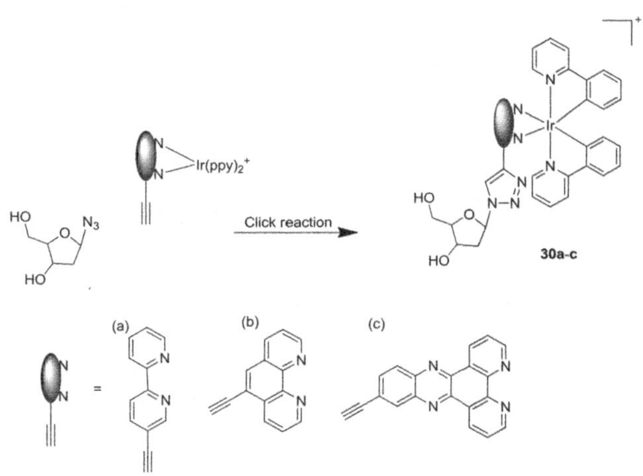

Fig. 9 Synthesis scheme for Ir (III) nucleoside

Fig. 10 Unnatural triazolyl nucleoside

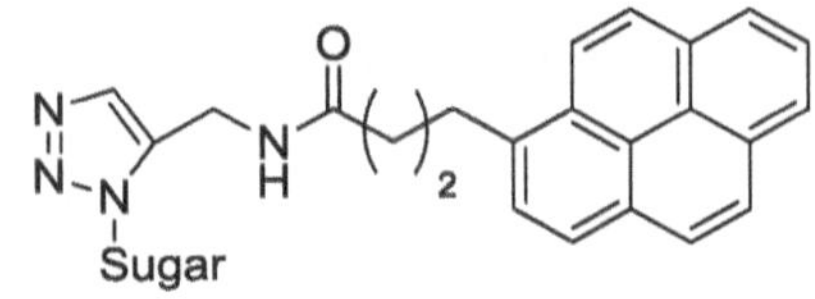

31 $^{TBy}B_{Do}$

Further, Bag et al. [42], reported the Förster resonance energy transfer (FRET) process in two chimeric DNA duplexes in which an unnatural nucleoside ($^{Tphen}B_{Do}$) was paired against a fluorescently labeled unnatural nucleoside ($^{oxopy}U$ or Per U). This pairing was not disturbing its B-DNA conformation and stability of the duplexes (Fig. 11).

Ozols et al. [43] developed a novel N^6-methyl-2-(1,2,3-triazol-1-yl)-2′-deoxyadenosine as a novel fluorescent triazolyl adenosine derivative. This modified nucleoside was further used for the synthesis of $GA^{Oz}C$ trinucleotide, which showed a moderate stoke shift and quantum yield (**32**). Kovalovs et al. [44], synthesized N^6-substituted-2-(1,2,3-triazol-1-yl)-adenine nucleoside by following three step synthetic approach. The fluorescence studies of the synthesized compounds (**33**) showed that they possess useful level of fluorescence. Ingale and Seela [45] synthesized fluorescent nucleoside and oligonucleoside with triazole linkers and

Fig. 11 Schematic representation of FRET process

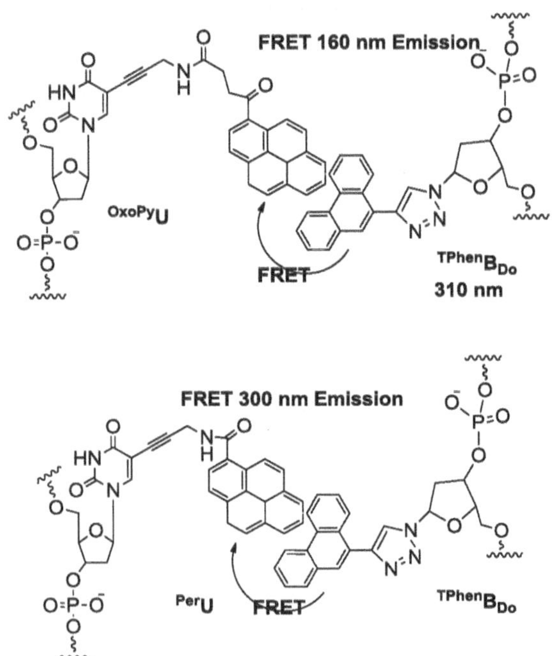

then studied the effect of these linkers on DNA duplex stability and fluorescence properties (**34** and **35**). Kosiova et al. [46]. synthesized coumarin-nucleoside conjugates as **36** and determined their fluorescence properties. Kavoosi et al. [47] recently synthesized 8-(1,2,3-triazol-1-yl)-7-deazapurine nucleosides (**37**) by using cycloaddition reactions of azide-alkyne and direct functionalization. These compounds showed moderate quantum yields and a large stoke shifts.

Seela et al. [48] synthesized coumarin-triazole conjugated nucleoside **38** and studied their fluorescence properties. Haque et al. [49] synthesized coumarin modified pyrimidine nucleoside having triazole moiety **39** and incorporated into oligonucleoside for DNA fluorescent tagging and for photoswitchable application (Fig. 12).

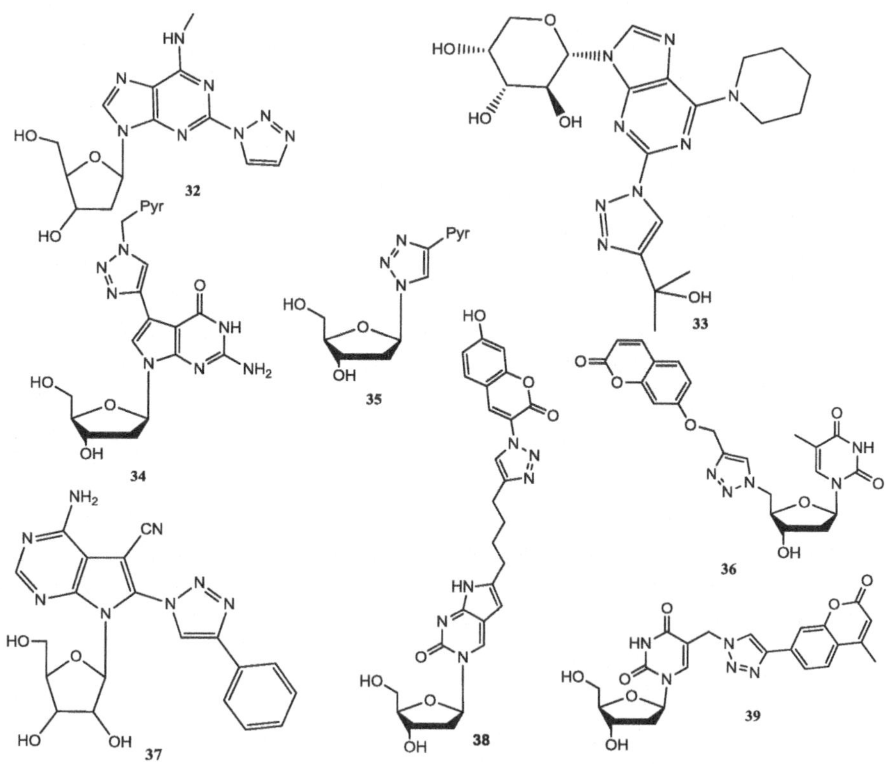

Fig. 12 Triazolyl nucleosides with various photophysical applications

2.9 *Miscellaneous*

This category includes group of recently synthesized triazolyl nucleosides with synthetic importance or few other characteristics, which cannot be accommodated in above all categories. Lopes et al. [50]. designed the selective 1,2,3-triazolehetrocycle N^2 arylation in 1-β-D-ribofuranosyl-2*H*-1,2,3-triazole substrate. The catalyst and the ligand system were optimized for this transformation which helps in the selective functionalization of triazole nucleosides (Fig. 13).

Carmen et al. [51] synthesized *C*-nucleosides, structurally similar to ribavarin, a nucleoside possessing broad spectrum antiviral properties. Synthesis of these compounds was based on an indium mediated alkynylglycosylation, followed by triazole ring formation. Primary and secondary amines were used for the amidation of these compounds (**40**).

Bag et al. [52], synthesized unnatural triazolyl nucleoside and studied the stabilization of DNA duplex containing these nucleosides. They reported that strong stacking propensity, large surface area and polarizability of triazolylphenanthrene helps in the thermal stabilization of $^{Tphe}B_{Do\text{-}\phi}$ duplex were comparable to that of natural AT pair (Fig. 14).

Novosjolova et al. [53], synthesized 2 and 6-alkyl/aryl thiopurine nucleoside having 1,2,3-triazole moiety (**41**, Fig. 15).

Upadhyay et al. [54] synthesized triazolyl fused quinolines with *N*-glycosyl substituents as unnatural nucleosides (**42**). Kumar et al. [55], synthesized oligonucleotides having triazole linkage between *C*-5 position of the uracil nucleobase and a benzenesulphonamide group. Among the studied oligonucleotides, the most stable DNA:RNA duplex was the one having four consecutive triazole linked sulphonamide substituted phenyl were in the major groove (**43**). Novosjolova et al. [56], designed and standardized reaction condition to synthesize the nucleoside analogs with 6-alkyl/arylthio-2-(4-alkyl/aryl-1-*H*-1,2,3-triazol-1-yl)purine bases (**44**). Paritala et al. [57], reported the microwave assisted solid phase coupling of nucleosides, including triazole containing nucleosides. They used mild cleavage conditions to release nucleoside derivatives from its solid support (aminomethylated polystyrene resin) (**45**). Novosjolova et al. [58], further synthesized

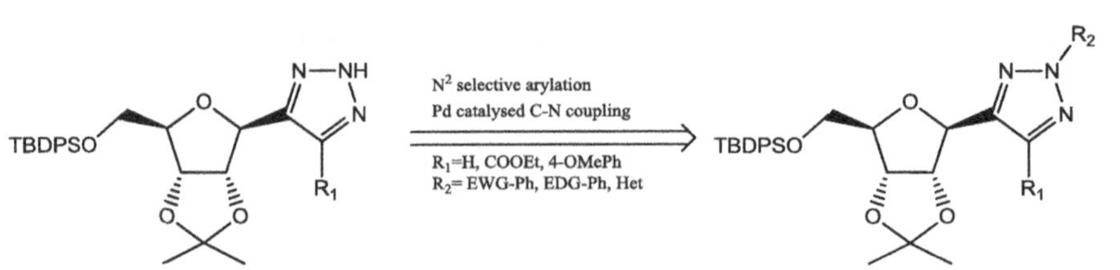

Fig. 13 Schemetic representation of selective functionalization of triazole nucleosides

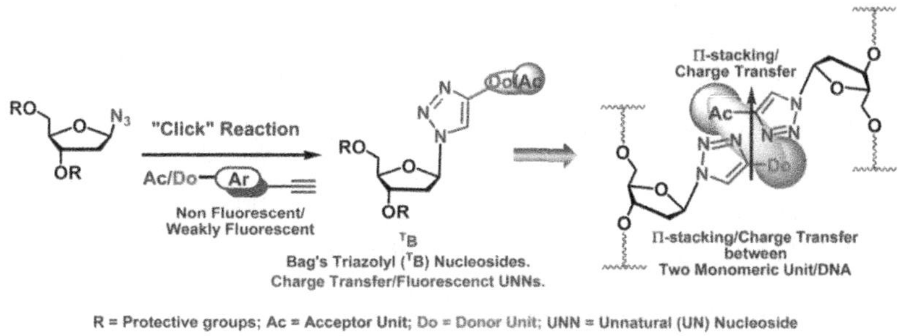

I

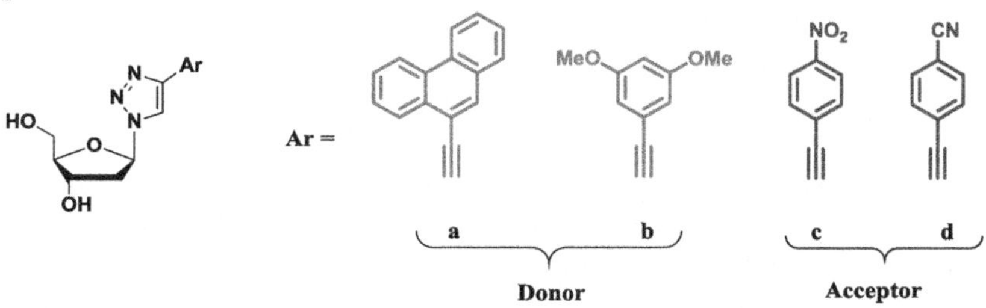

II

Fig. 14 **I** Schematic presentation of duplex stabilization by unnatural triazolyl nucleosides; **II** structures of alkynes and nucleosides that were incorporated into DNA (Reprinted with permission from American Chemical Society, © 2013 ACS publication)

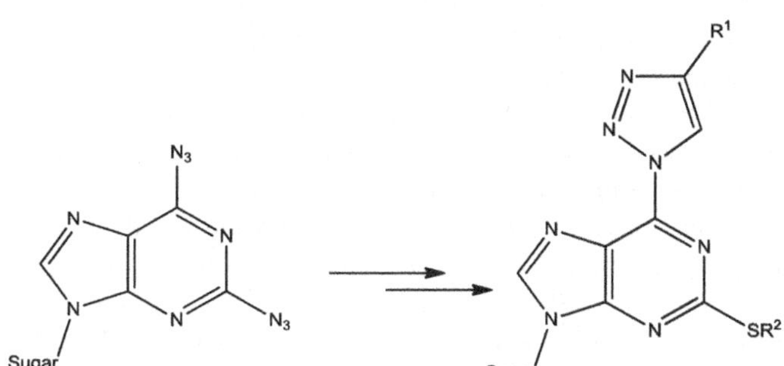

41

Fig. 15 Schemetic representation of synthesis of thiopurine nucleoside

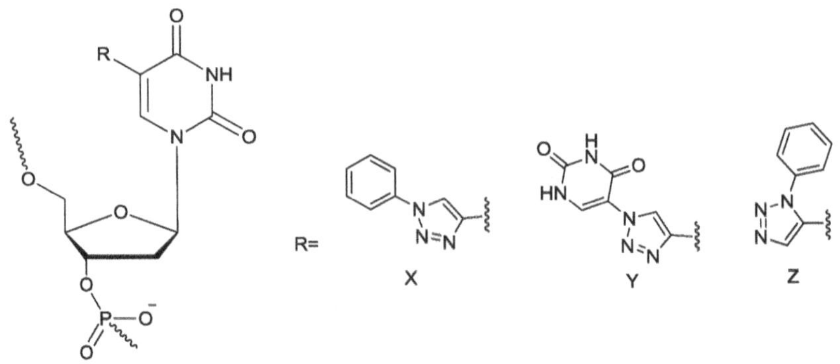

47

Fig. 16 *C*-5 thymine functionalized triazolyl nucleosides

2,6-bistriazolylpurine arabnonucleoside by using nucleophillic substitution reaction of 1,2,3-triazole moiety with pyrrolidine. NMR and X-ray studies were used to confirm the anomeric configuration and pyranose form of arabinose. UV/visible and emission spectra of the synthesized compound were studied (**46**). Hornum et al. [59], synthesized thymidine analogues with *C*-5 triazolyl moiety and introduced two of these new derivatives (**47**) in DNA:DNA and DNA:RNA duplexes in order to determine their thermal effects by changing 1,4-triazoles with 1,5-triazoles and by changing phenyl substituent (apolar) with uracil-5-yl substitutent (polar) (Fig. 16).

Rad et al. [60] did a highly efficient synthesis of few carboacyclic nucleosides with 1,2,3-triazole moiety (**48**) by using click reaction of *N*-propargylated nucleobases and azido alcohols in the presence of Cu/aminoclay/reduced graphene oxide nanohybrid as a nanocatalyst. They reported that this nanohybrid was stable and efficient heterogenous nanocatalyst and can be prepared easily and also restored from the reaction by simple filtration for reuse, without any significant decrease in activity even after many consecutive trials. Elayadi et al. [61], synthesized 1,4-disubstituted 1,2,3-triazolonucleosides using $CuSO_4$ and KI as copper (I) catalyst sources (**49**). Prasad, et al. synthesized 3′-deoxy-3′-(4-substituted-triazol-1-yl)-5-methyluridine and 2′-triazolo nucleoside conjugates **50** [62] and **51** [63] and evaluated their biological activity. Jawalekar et al. [64] synthesized coumarin conjugated nucleosides **52** by using 1,3-dipolar cycloaddition reaction (Fig. 17).

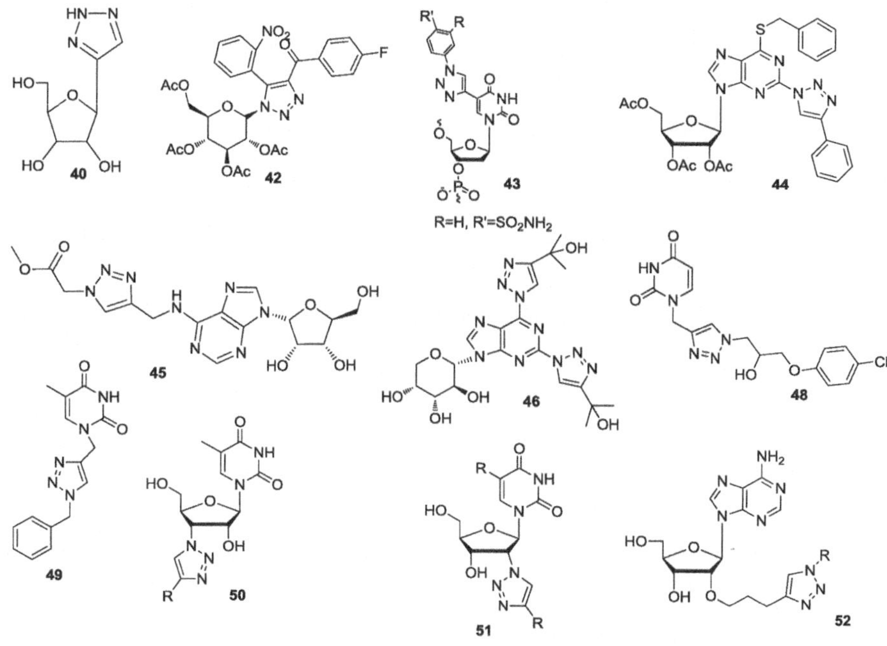

Fig. 17 Various synthesized triazolyl nucleosides

3 Conclusion

Recent reports on triazolyl nucleosides showed that this is a growing area of research, as these group of compounds play significant roles for pharmacological applications, photophysical applications and several other studies. This review thus gives a brief insight of synthesis, application and importance of variously modified recently reported triazolyl nucleosides.

References

1. (a) Matsuda A, Sasaki T (2004) Antitumor activity of sugar-modified cytosine nucleosides. Cancer Sci 95:105; (b) Vester B, Wengel J (2004) LNA (Locked Nucleic Acid): High-Affinity targeting of complementary RNA and DNA. Biochemistry 43:13233; (c) Russ P, Schelling P, Scapozza L, Folkers G, De Clercq E, Marquez VE (2003) Synthesis and biological evaluation of 5-substituted derivatives of the potent antiherpes agent (north)-Methanocarbathymine. J Med Chem 46:5045; (d) Agrofoglio LA, Challand SR, Huryn DM, Okabe M (1992) AIDS driven nucleoside chemistry. Chem Rev 92:1745
2. (a) Aartsma-Rus A (2012) Overview on AON design. Methods Mol Biol 867:117; (b) Yamamoto T, Harada-Shiba M, Nakatani M, Wada S, Yasuhara H, Narukawa K, Sasaki K, Shibata MA, Torigoe H, Yamaoka T, Imanishi T, Obika S (2012) Cholesterol-lowering action of BNA-based Antisense Oligonucleotides targeting

PCSK9 in Atherogenic Diet-induced Hypercholesterolemic Mice. Mol Ther Nucleic Acids 1:22; (c) Bennett CF, Swayze EE (2010) RNA targeting therapeutics: molecular mechanisms of antisense Oligonucleotides as a therapeutic platform. Annu Rev Pharmacol Toxicol 50:259. (d) Kole R, Krainer AR, Altman S (2012) RNA therapeutics: beyond RNA interference and antisense oligonucleotides. Nat Rev Drug Discov 11:125; (e) Dagle JM, Weeks DL, Walder JA, (1991) Pathways of degradation and mechanism of action of antisense oligonucleotides in Xenopus laevis embryos. Antisense Res Dev 1:11; (f) Eder PS, Devine RJ, Dagle JM, Walder JA (1991) Substrate specificity and kinetics of degradation of antisense oligonucleotides by a 3′ exonuclease in plasma. Antisense Res Dev 1:141

3. Mickelfield J (2001) Backbone modification of nucleic acid: synthesis, structure and therapeutic applications. Curr Med Chem 8:1157
4. Sipa K, Sochacka E, Kazmierczak-Baranska J, Maszewska M, Janicka M, Nowak G, Nawrot B (2007) Effect of base modifications on structure, thermodynamic stability, and gene silencing activity of short interfering RNA. RNA 13:1301
5. Prakash TP (2011) An overview of sugar-modified Oligonucleotides for antisense therapeutics. Chem Biodivers 8:1616
6. Mulamba GB, Hu A, Azad RF, Anderson KP, Coen DM (1998) Human cytomegalovirus mutant with sequence-dependent resistance to the phosphorothioate oligonucleotide fomivirsen (ISIS 2922). Antimicrob Agents Chemother 42:971
7. Merki E, Graham MJ, Mullick AE, Miller ER, Crooke RM, Pitas RE, Witztum JL, Tsimikas S (2008) Antisense oligonucleotide directed to human apolipoprotein B-100 reduces lipoprotein (a) levels and oxidized phospholipids on human apolipoprotein B-100 particles in lipoprotein (a) transgenic mice. Circulation 118:743
8. Huisgen R (1963) Kinetics and mechanism of 1,3-dipolar cycloadditions. Angew Chem 75:604
9. Meldal M, Tornoe CW (2008) Cu-catalyzed azide-alkyne cycloaddition. Chem Rev 108:2952
10. Kolb HC, Sharpless KB (2003) The growing impact of click chemistry on drug discovery. Drug Discov Today 8:1128; (b) Moses JE, Moorhouse AD (2007) The growing applications of click chemistry. Chem Rev 36:1249
11. (a) Lolk L, Pohlsgaard J, Jepsen AS, Hansen LH, Nielsen H, Steffansen SI, Sparving L, Nielsen AB, Vester B, Nielsen P (2008) A click chemistry approach to pleuromutilin conjugates with nucleosides or acyclic nucleoside derivatives and their binding to the bacterial ribosome. J Med Chem 51:4957 (b) Moustafa AH, Haggam RA, Younes ME, El Ashry ESH (2006) The synthesis of triazolothiadiazines and thiadiazoles from 1,2-bis-(4-amino-5-mercapto-1,2,4-triazol-3-yl)-ethanol and ethane. Phosphorus, Sulfur Silicon Relat Elem 181:2361
12. Moustafa AH, El-Sayed HA, Haikal AEFZ, El Ashry ESH (2011) Synthesis of acyclovir and HBG analogues having Nicotinonitrile and its 2-methyloxy 1,2,3-triazole. Nucleosides Nucleotides Nuclic Acids 30:340
13. (a) El Sadek MM, Abd El-Dayem NS, Hassan SY, Mostafa MA, Yacout GA (2014) Antioxidant and antitumor activities of new synthesized aromatic *C*-nucleoside derivatives. Molecules 19:5163 (b) Wang M, Xia Y, Fan Y, Rocchi P, Qu F, Iovanna JL, Peng L (2010) A novel arylethynyltriazole acyclonucleoside inhibits proliferation of drug-resistant pancreatic cancer cells. Bioorg Med Chem Lett 20:5979
14. Wu J, Yu W, Fu L, He W, Wang Y, Chai B, Song C, Chang (2013) 2′-deoxy-2′-fluoro-4′-triazole cytidine nucleosides as potent antiviral agents. Eur J Med Chem 63:739
15. Alaoui S, Dufies M, Driowya M, Demange L, Bougrin K, Robert G, Auberger P, Pagès G, Benhida R (2017) Synthesis and anti-cancer activities of new sulfonamides 4-substituted-triazolyl nucleosides. Bioorg Med Chem Lett 27:1989–1992
16. Amdouni H, Robert G, Driowya M, Furstoss N, Métier C, Dubois A, Dufies M, Zerhouni M, Orange F, Gervais SL, Bougrin K, Martin AR, Auberger P, Benhida R (2017) In Vitro and in Vivo evaluation of fully substituted (5-(3-ethoxy-3-oxopropynyl)-4-(ethoxycarbonyl)-1,2,3-triazolyl-glycosides as original nucleoside analogues to circumvent resistance in myeloid malignancies. J Med Chem 60:1523–1533

17. Bodnár B, Mernyák E, Wölfling J, Schneider G, Herman BE, Szécsi M, Sinka I, Zupkó I, Kupihár Z, Kovács L (2016) Synthesis and biological evaluation of triazolyl 13α-estrone–nucleoside. Molecules 21:1212
18. Giofrè SV, Romeo R, Carnovale C, Mancuso R, Cirmi S, Navarra M, Garozzo A, Chiacchio MA (2015) Synthesis and biological properties of 5-(1H-1,2,3-Triazol-4-yl) isoxazolidines: a new class of *C*-nucleosides. Molecules 20
19. Romeo R, Carnovale C, Giofrè SV, Chiacchio MA, Garozzo A, Amata E, Romeo G, Chiacchio U (2015) C-5′-triazolyl-2′-oxa-3′-aza-4′a-carbanucleosides: synthesis and biological evaluation. Beilstein J Org Chem 11:328–334
20. Zhang Q, He P, Zhou G, Gu Y, Fu T, Xue D, Liu HM (2013) Synthesis and biological evaluation of 2,6-dichloropurine bicyclonucleosides containing a triazolyl-carbohydrate moiety. Carbohyd Res 382:65–70
21. Lakshman MK, Kumar A, Balachandran R, Day BW, Andrei G, Snoeck R, Balzari J (2012) Synthesis and biological properties of *C*-2 triazolylinosine derivatives. J Org Chem 77: 5870–5883
22. Ouahrouch A, Taourirte M, Schols D, Snoeck R, Andrei G, Engels JW, Lazrek HB (2016) Design, synthesis, and antiviral activity of novel ribonucleosides of 1,2,3-triazolylbenzyl-aminophosphonates. Arch Pharm Chem Life Sci 349:30–41
23. Rashad AE, Shamroukh AH, El-Sayed AH, Micky JA, Marsok NA, Abdel-Megeid FME (2016) Some 1,2,4-triazole derivatives: synthesis and antiviral evaluation. Der Pharma Chemica 8(15):75–81
24. Vernekar, Kumar SV, Li Q, Jeana Z, Geraghty RJ, Zhengqiang W (2014) Synthesis and antiviral evaluation of 4′-(1,2,3-triazol-1-yl)thymidines. Med Chem Comm 5(5):603–608
25. Jamal K, Moha T, Christian G, Ivan K, Joachim EW (2013) Microwave-assisted click chemistry for nucleoside functionalization: useful derivatives for analytical and biological applications. Synthesis 45(3):396–405
26. Elayadi H, Smietana M, Vasseur JJ, Balzarini J, Lazrek HB (2014) Synthesis of 1,2,3-triazolyl nucleoside analogs as potential anti-influenza A (H3N2 Subtype) virus agents. Arch Pharm Chem Life Sci 347:134–141
27. Zhou L, Amer A, Korn M, Burda R, Balzarini J, De Clercq E, Kern ER, Torrence PF (2005) Synthesis and antiviral activities of 1,2,3-triazole functionalized thymidines: 1,3-dipolar cycloaddition for efficient regioselective diversity generation. Antivir Chem Chemother 16:375
28. Montague A, Roy V, Balzarini J, Snoeck R, Andrei G, Agrofoglio LA (2011) Synthesis of new C5-(1-substituted-1,2,3-triazol-4 or 5-yl)-2′-deoxyuridines and their antiviral evaluation. Eur J Med Chem 46:778
29. Vernekar SKV, Qiu L, Zacharias J, Geraghty RJ, Wang Z (2014) Synthesis and antiviral evaluation of 4′-(1,2,3-triazol-1-yl)thymidines. Med Chem Commun 5:603
30. Olomola TO, Klein R, Lobb KA, Sayed Y, Kaye PT (2010) Towards the synthesis of coumarin derivatives as potential dual-action HIV-1 protease and reverse transcriptase inhibitors. Tetrahedron Lett 51:6325
31. Hwu JR, Lin SY, Tsay SC, De Clercq E, Leyssen P, Ney J (2011) Coumarin-Purine ribofuranoside conjugates as new agents against hepatitis C virus. J Med Chem 54:2114
32. Xavier NM, Lucas SD, Jorda R, Schwarz S, Loesche A, Csuk R, Oliveira MC (2015) Synthesis and evaluation of the biological profile of novel analogues of nucleosides and of potential mimetics of sugar phosphates and nucleotides. Synlett 26(19):2663–2672
33. Mahony GO, Svensson S, Sundgren A, Grotli M (2008) Synthesis of 2′-([1,2,3]triazol-1-yl)-2′-deoxyadenosines. Nucleosides Nucleotides Nucleic Acids 27:449
34. Daligaux P, Pomel S, Leblanc K, Loiseau PM, Cavé C (2016) Simple and efficient synthesis of 5′-aryl-5′-deoxyguanosine analogs by azide-alkyne click reaction and their antileishmanial activities. Mol Divers 20:507–519
35. Reddy PV, Saquib M, Mishra NN, Shuklab PK, Shaw AK (2014) Stereoselective synthesis of tetrahydrofuranyl 1,2,3-triazolyl *C*-nucleoside analogues by 'click' chemistry and investigation of their biological activity. ARKIVOC (iv):170–182

36. Tuberculosis; World Health Organization report 2016. http://www.who.int/tb/research/en
37. GreenFacts. Scientific facts on drug-resistant Tuberculosis. http://www.greenfacts.org/en/tuberculosis/l-2/1-mdr-tb-xdr.htm
38. Poecke SV, Munier-Lehmann MH, Helynck O, Froeyen M, Calenbergh SV (2011) Synthesis and inhibitory activity of thymidine analogues targeting *Mycobacterium tuberculosis* thymidine monophosphate kinase. Bioorg Med Chem 19:7603
39. Tampure AA, Pawar MG, Srivatsan SG (2013) Fluorescent nucleoside analogs: probes for investigating nucleic acid structure and function. Isr J Chem 53:366–378
40. Passays J, Rubay C, Marcelis L, Benjamin E (2017) Synthesis and photophysical properties of triazolyl irIII nucleosides. Eur J Inorg Chem 623–629
41. Bag SS, Talukdar S, Das SK, Pradhan MK, Mukherjee S (2016) Donor/acceptor chromophores-decorated triazolyl unnatural nucleosides: synthesis, photophysical properties and study of interaction with BSA. Org Biomol Chem 14:5088
42. Bag SS, Das SK, Pradhan MK, Jana S (2016) Hybridization accompanying FRET event in labeled natural nucleoside-unnatural nucleoside containing chimeric DNA duplexes. J Photochem Photobiol, B 162:669–673
43. Ozols K, Cırule D, Novosjolova I, Stepanovs D, Liepinsh E, Bizdena E, Turks M (2016) Development of N^6-methyl-2-(1,2,3-triazol-1-yl)-2'-deoxyadenosine as a novel fluorophore and its application in nucleotide synthesis. Tetrahedron Lett 57:1174–1178
44. Kovalovs A, Novosjolova I, Bizdena E, Bizane I, Skardziute L, Kazlauskas K, Jursenas S, Turks M (2013) 1,2,3-triazoles as leaving groups in purine chemistry: a three-step synthesis of N^6-substituted-2-triazolyl-adenine nucleosides and photophysical properties thereof. Tetrahedron Lett 54:850–853
45. Ingale SA, Seela F (2014) Nucleoside and oligonucleotide pyrene conjugates with1,2,3--triazolyl or ethynyl linkers: synthesis, duplex stability, and fluorescence changes generated by the DNA-dye connector. Tetrahedron 70:380–391
46. Kosiova I, Kovackova S, Kois P (2007) Synthesis of coumarin–nucleoside conjugates via huisgen 1,3-dipolar cycloaddition. Tetrahedron 63:312–320
47. Kavoosi S, Rayala R, Walsh B, Barrios M, Gonzalez WG, Miksovska J, Mathivathanan L, Raptis RG, Wnuk SF (2016) Synthesis of 8-(1,2,3-triazol-1-yl)-7-deazapurine nucleosides by azide–alkyne click reactions and direct C-H bond functionalization. Tetrahedron Lett 57:4364–4367
48. Seela F, Sirivolua VR (2008) Pyrrolo-dC oligonucleotides bearing alkynyl side chains with terminal triple bonds: synthesis, base pairing and fluorescent dye conjugates prepared by the azide–alkyne “click” reaction. Org Biomol Chem 6:1674–1687
49. Haque MM, Sun H, Liu S, Wang Y, Peng X (2014) Photo-Switchable DNA interstrand cross-link formation by a coumarin-modified nucleotide. Angew Chem Int Ed 53:7001
50. Lopes AB, Wagner P, Alves Rodrigo, de Souza OM, Germain NL, Uziel J, Bourguignon JJ, Schmitt M, Miranda LSM (2016) Functionalization of 2*H*-1,2,3-Triazole *C*-Nucleoside Template via N^2 Selective Arylation. J Org Chem 81:4540–4549
51. Carmen S, Michael DS, Simon G, Leandro MSM, Regis G, Angelique F, Florian G, Jacques U, Nadege LG (2017) Synthesis of *C*-Ribosyl-1,2,3-triazolyl Carboxamides. Synthesis 49(9):1993–2002
52. Bag SS, Talukdar S, Matsumoto K, Kundu R (2013) Triazolyl donor/acceptor chromophore decorated unnatural nucleosides and oligonucleotides with duplex stability comparable to that of a natural adenine/thymine pair. J Org Chem 78:278
53. Novosjolova I, Bizdēna E, Turks M (2015) Synthesis of novel 2- and 6-alkyl/arylthiopurine derivatives. Phosphorus Sulfur Silicon 190:1236–1241
54. Upadhyaya K, Ajay A, Mahar R, Pandey R, Kumar B, Shukla SK, Tripathi RP (2013) A strategy to access fused triazoloquinoline and related nucleoside analogues. Tetrahedron 69:8547–8558
55. Kumar P, Chandak N, Nielsen P, Sharma PK (2012) Sulfonamide bearing oligonucleotides: simple synthesis and efficient RNA recognition. Bioorg Med Chem 20:3843–3849

56. Novosjolova I, Bizdena E, Turks M (2013) Application of 2,6-diazidopurine derivatives in the synthesis of thiopurine nucleosides. Tetrahedron Lett 54:6557–6561
57. Paritala H, Suzuki Y, Carroll KS (2013) Efficient microwave-assisted solid phase coupling of nucleosides, small library generation, and mild conditions for release of nucleoside derivatives. Tetrahedron Lett 54:1869–1872
58. Novosjolova I, Bizdēna E, Belyakov S, Turks M (2013) The Synthesis and X-ray studies of 6-pyrrolidinyl-2-triazolyl purine arabinonucleoside. Mater Sci Appl Chem 28:39–44
59. Hornum M, Djukina A, Sassnau AK, Poul Nielsen P (2016) Synthesis of new *C*-5-triazolyl-functionalized thymidine analogs and their ability to engage in aromatic stacking in DNA : DNA and DNA : RNA duplexes. Org Biomol Chem 14:4436
60. Rad MNS, Behrouz S, Hoseini SJ, Nasrabadi H, Zare MS (2015) Copper/Graphene/Clay Nanohybrid: a highly efficient heterogeneous nanocatalyst for the synthesis of novel 1,2,3-triazolyl carboacyclic nucleosides via 'Click' Huisgen 1,3-dipolar cycloaddition. Helvetica Chimica Acta 98:1210
61. Elayadi H, Lazrek HB (2015) CuSO4/KI as catalyst for the synthesis of 1,4-disubstituted-1,2,3-triazolo-nucleosides. Nucleosides Nucleotides Nucleic Acids 34:433–441
62. Arya A, Mathur D, Tyagi A, Kumar R, Kumar V, Olsen CE, Saxena RK, Prasad AK (2013) Chemoenzymatic synthesis of 3′-deoxy-3′-(4-substituted-triazol-1-yl)-5-methyluridine. Nucleosides Nucleotides Nucleic Acids 32:646
63. Mathur D, Rana N, Olsen CE, Parmar VS, Prasad AK (2015) Cu(I)-catalyzed efficient synthesis of 2′-triazolo-nucleoside conjugates. J Heterocyclic Chem 52:701
64. Jawalekar AM, Meeuwenoord N, Cremers J (Sjef) GO, Overkleeft HS, Marel GA van der, Rutjes FPJT, Floris L van D (2008) Conjugation of nucleosides and oligonucleotides by [3+2] cycloaddition. J Org Chem 73:287–290

In House Laboratory Method Validation and Uncertainty Determination of 28 Pesticides in Spinach by Gas Chromatography Using Electron Capture (ECD) and Mass Spectrometric (MS) Detector

Sudeep Mishra, Neelam Richhariya, Samsul Alam and Lalitesh K. Thakur

Abstract Study conducted to develop and validate laboratory method using QuEChERS method. QuEChER method was used for the determination of 28 pesticide residues belongs to organochlorine, synthetic pyrethroids and herbicides in Spinach. Sample was extracted by ethyl acetate. Removal of moisture using from sample extract by anhydrous sodium sulphate and magnesium sulphate. Cleaning of matrix and other interferences was done by Primary Secondary Amine (PSA), C18 and activated charcoal. Agilent 7890B model GC-ECD with 7693 auto sampler and 7890A-Agilent Technologies GC with 5975C inert XL EI/CI MSD triple axis mass detector with DB-5MS fused silica capillary column were used for analysis of samples taken for study. Each pesticide spiking concentration was 0.05 ppm level in three replicate whose recovery lies between Codex permissible limit i.e. 70–120%. Relative standard deviation (RSD) < 15%. Codex norms were followed for repeatability and reproducibility check of the instrument. For Linearity check, taking area of lambda cyhalothrin against five different concentration levels of 0.003, 0.010, 0.100, 0.70, 1.00 mg/kg, calibration curve was plotted with regression co-efficient (r^2) 0.993. Total uncertainty or combined uncertainty comprise of three main independent uncertainties arise due to weighing of standards, purity of the standard and repeatability of instruments. Expanded uncertainty is twice of the combined uncertainty which lies in three ranges viz., (a) $\leq 10\%$ (b) 11–15% and (c) 15–20%. All pesticides taken for study were in the range (a) i.e. $\leq 10\%$.

S. Mishra · S. Alam · L. K. Thakur (✉)
Institute of Pesticide Formulation Technology, (Ministry of Chemicals & Fertilizers, Govt. of India), Sector-20, Udyog Vihar, Opp. Ambience Mall, Gurugram 122016, Haryana, India
e-mail: thakurlk1964@rediffmail.com

N. Richhariya
Govt. Indra College Satna, Satna 485001, Madhya Pradesh, India

V. S. Parmar et al. (eds.), *Green Chemistry in Environmental Sustainability and Chemical Education*, https://doi.org/10.1007/978-981-10-8390-7_17

Keywords Spinach · Pesticides · LOD · LOQ · QuEChERS · Uncertainty

1 Introduction

For safe use of pesticides in house hold and agriculture purposes, pesticide residue analysis plays important role in making end user aware about safe use of pesticides. Pesticide analysis depends upon quality parameters such as selectivity, repeatability, precision and sensitivity to meet the need for any particular analysis. Analytical methods to measure accurately small amounts of pesticides residues present in foodstuffs are very much required. To determine multiresidue, numbers of analytical methods are in common practices. The analytical techniques for all types of analyses and matrices monitored have also advanced for various types of sample preparation techniques. For analysis of pesticide residues in food matrices, Anastassiades et al. [1] described the quick, easy, cheap, effective, rugged, low solvent consumption, wide pesticide range and safe (QuEChERS) method. For including QuEChERS method as an official method for Association of Analytical Communities (AOAC) Official Method 2007.01., it has been validated for hundreds of pesticide residues in different types of matrices. The advantages of QuEChERS method over most traditional methods of analysis are high recoveries (greater than 85%) which are achieved for a wide polarity and volatility range of pesticides, including notoriously difficult analytes.

The present study is undertaken with the objective to validate method which scrutinize and quantify the residue levels of some most frequently used pesticides on spinach leaves. For the study of persistence of pesticide in spinach pesticides, a routine multi-residue pesticide analysis method should be followed. Gas chromatography (GC) enabled with an electron-capture detector (ECD), Flame Photometric Detector (FPD), Nitrogen Phosphorus Detector (NPD) or mass spectrometer (MS) has been the analytical technique more frequently applied to the analysis of pesticides in food for many years [1–5].

2 Result and Discussion

Laboratory method was validated to target 28 organochlorine pesticides commonly used on spinach matrix. For establishing method performance criteria, single laboratory validation and pesticide residue analysis were carried as per European guidelines.

2.1 Pesticides Taken for Study

28 pesticides which are commonly used in India viz. Alpha-HCH, Beta-HCH, Gamma HCH, Delta HCH, Alachlor, Aldrin, Dicofol, Pendimethalin, O,P DDE, Alpha-Endosulphan, Heptachlor, P,P DDE, Endosulphan Sulphate, Dialdrin, O,P DDD, Beta-Endosulphan, P,P DDD, O,P DDT, P,P DDT, Bifenthrin, Fenpropathrin, Lambda Cyhalothrin, Beta Cyfluthrin, Cypermethrin, Fenvalarate, Fluvalinate, Deltamethrin are considered for study in spinach matrix. Out of 28 pesticides taken Endosulphan, BHC and DDT isomers are highly persistent in environment. Preliminary screening was done to find out the best chromatographic techniques which can be used in terms of LOD/LOQ, response and peak shape. All 28 pesticides gave better response for ECD detector, which was used for quantification and confirmation by GC-Mass detector. Figure 1 shows standard mix chromatogram of 28 pesticides of 0.05 ppm concentration.

2.2 Limit of Quantification (LOQ) and Limit of Detection (LOD)

LOD and LOQ will be measured by using EPA method as it is simple, easy and practical to implement [11–13]. To measure the LOD, peak to peak noise of blank matrix (spinach) at or around the retention time of individual pesticides chromatogram of standard mix is noted and averaged for three replicates. Concentration of the individual pesticide is calculated (in µg/g) from the matrix spiked chromatogram which could produce the signal equal to three times of blank matrix. Figure 2 shows GC-ECD chromatogram of spinach blank matrix. LOQ is calculated by multiplying the LOD value by factor 3 round of two decimal place value.

Measured method are listed in Table 1 shows LOD, LOQ, MRLs and ADI values. With the present method, in case of spinach, all 28 pesticides for study resulted LOD and LOQ values which are below MRL. It always happens that

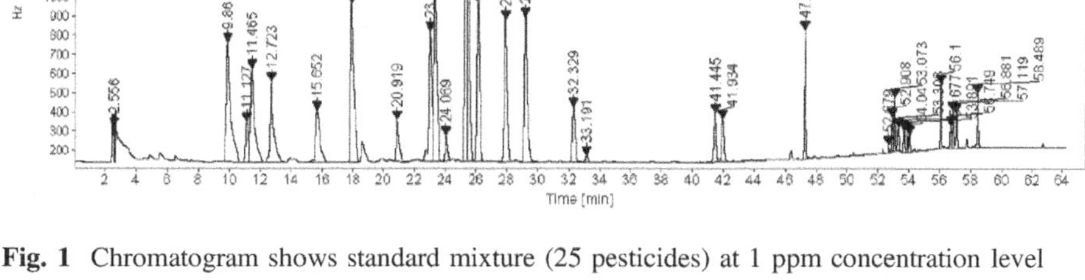

Fig. 1 Chromatogram shows standard mixture (25 pesticides) at 1 ppm concentration level

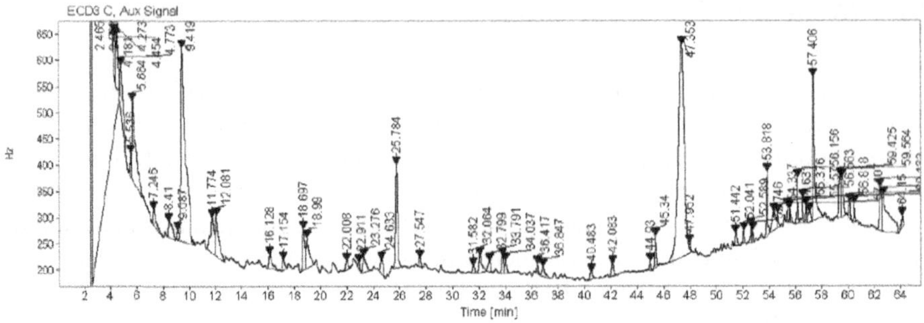

Fig. 2 GC-ECD chromatogram of spinach matrix (blank)

Table 1 Shows class, retention time (RT), limit of detection (LOD), limit of quantification (LOQ), maximum residues limits (MRL) and acceptable daily intake (ADI) of organochlorine, synthetic pesticide and herbicides from spiked spinach matrix by GC-ECD

S. No.	Pesticides	Class	RT	LOD (mg/kg)	LOQ (mg/kg)	MRL (mg/kg)	ADI (mg/kg/day)
1	Alpha-HCH	Organochlorine	9.80	0.003	0.009	1.00	0.004
2	Dicofol	Organochlorine	10.80	0.003	0.010	1.00	0.002
3	Beta-HCH	Organochlorine	11.25	0.004	0.012	1.00	0.0005
4	Gamma HCH	Organochlorine	11.57	0.003	0.009	1.00	0.0005
5	Delta HCH	Organochlorine	12.86	0.002	0.006	1.00	0.0005
6	Heptachlor	Herbicide	15.78	0.008	0.024	NA	0.003
7	Alachlor	Herbicide	15.87	0.01	0.030	NA	0.0005
8	Aldrin	Organochlorine	18.05	0.005	0.015	0.1	0.0001*
9	Pendimethalin	Herbicide	21.06	0.005	0.015	NA	0.002
10	O,P DDE	Organochlorine	23.21	0.009	0.027	3.50	0.003
11	Alpha-Endosulphan	Organochlorine	23.52	0.004	0.012	2	0.01
12	Butachlor	Herbicide	24.22	0.003	0.009	NA	0.003
13	Dialdrin	Organochlorine	25.53	0.010	0.030	NA	NA
14	P,P DDE	Organochlorine	25.68	0.006	0.018	3.50	0.005
15	O,P DDD	Organochlorine	26.30	0.008	0.024	3.50	0.02
16	P,P DDT	Organochlorine	26.40	0.004	0.012	3.50	NA
17	Beta-Endosulphan	Organochlorine	28.14	0.005	0.015	2	0.01
18	P,P DDD	Organochlorine	29.47	0.005	0.015	3.50	NA
19	O,P DDT	Organochlorine	29.72	0.006	0.018	3.50	NA
20	Endosulphan Sulphate	Organochlorine	32.60	0.004	0.012	NA	NA
21	Bifenthrin	Synthetic Pyrethroids	41.72	0.007	0.021	NA	0.004
22	Fenpropathrin	Synthetic Pyrethroids	42.19	0.002	0.006	NA	0.03*

(continued)

Table 1 (continued)

S. No.	Pesticides	Class	RT	LOD (mg/kg)	LOQ (mg/kg)	MRL (mg/kg)	ADI (mg/kg/day)
23	Lambda Cyhalothrin	Synthetic Pyrethroids	47.44	0.01	0.03	NA	0.0005
24	Beta Cyfluthrin	Synthetic Pyrethroids	52.7–52.9	0.01	0.03	NA	0.002
25	Cypermethrin	Synthetic Pyrethroids	53.03–53.44	0.005	0.015	0.02	0.003
26	Fenvalarate	Synthetic Pyrethroids	56.23	0.005	0.015	2	0.003*
27	Fluvalinate	Synthetic Pyrethroids	56.9–57.2	0.01	0.03	NA	NA
28	Deltamethrin	Synthetic Pyrethroids	58.63	0.006	0.018	NA	0.02

ADI Acceptable daily intake; *MRL* maximum residue limit (PFA, 1954); *NA* not available

matrix effect is predominant for some pesticides. The quantification was done using matrix-matched standards prepared in matrix blank, to overcome variable matrix effect.

2.3 *Linearity*

The calibration curves of GC-ECD instrument were created at six levels i.e. (matrix matched) prepared by spiking pesticide mixture solution at different concentration levels in blank sample extract and injected in triplicate. Figure 3 shows calibration curve were plotted between an area of lambda cyhalothrin pesticide against concentration levels of 0.003, 0.010, 0.100, 0.70, 1.00 mg/kg with regression co-efficient (r^2) 0.993.

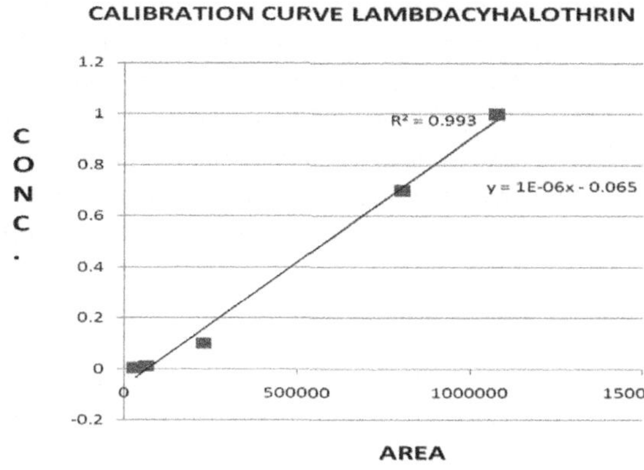

Fig. 3 Linearity curve of lambda cyhalothrin

2.4 Recovery

By recovery studies method trueness was assessed, using blank spinach matrices spiked at 0.05 ppm concentration level, prior to sample preparation and injection in three prepared replicates individually. Recovery, standard deviation and relative standard deviation (% RSD) were calculated (Table 3). As per SANCO guideline, recovery of spiking concentration should come within acceptable range between 70 and 120%. For routine measurement, specially optimized analytical methods should be applied for the analysis of pesticides present in spinach matrices. QuEChER extraction method is followed for recovery study of spiked pesticides from spinach matrix. High mean recovery between 90 and 100% were calculated for Dialdrin, Pendimethalin, Bifenthrin, Gamma HCH and Dicofol while less recovery (between 75 and 80) was calculated for O,P DDT and P,P DDD. Rest all pesticides recovery was calculated between 80 and 90%.

2.5 Selectivity

NIST library of Gas chromatography mass spectrometry was used to confirm the presence of pesticides taken for study. Spectra matched with reference spectra of NIST. Presence of pesticides was confirmed by their fragmentation pattern matches with reference spectra of NIST. Table 2 shows mass fragmentation pattern of each pesticide for conformation in GC-MS. Samples blank were also inspected in GC-ECD for the presence of interfering peaks near to the target retention time. Peaks closer to target peaks in blank samples were observed for all 28 pesticides taken for study.

2.6 Repeatability and Reproducibility

To study repeatability of the method, samples were analysed for recovery studies and repeated for three times are R1, R2 and R3. Mean (M), Standard Deviation (SD), and Relative Standard Deviation (RSD) of each pesticide were calculated as shown in Table 3. Precision was tested for both retention time and peak area of all pesticides taken for study by repeatable injections (n = 3) at low concentration level (0.003 ppm) standard solutions which was below 0.5% for all compounds. <20% was set as acceptance criteria as per SANCO requirements, for the target compounds and matrices.

Table 2 Fragmentation pattern of each pesticide taken for study

S. No.	Pesticide	Qualifier ions (m/z)		
1	Alachlor	188	160	–
2	Aldrin	277	263	261
3	Alpha-Endosulphan	241	195	–
4	Alpha-HCH	219	183	181
5	Beta Cyfluthrin	226	206	163
6	Beta-Endosulphan	241	237	195
7	Beta-HCH	219	183	181
8	Bifenthrin	181	166	–
9	Delta HCH	219	183	181
10	Dicofol	251	250	139
11	Fenpropathrin	349	265	181
12	Fenvalarate	419	167	125
13	Fluvalinate	252	250	–
14	Gamma HCH	219	183	181
15	Lambda Cyhalothrin	449	197	181
16	O,P DDT	237	235	212
18	P,P DDD	237	235	178
19	P,P DDE	318	226	–
20	P,P DDT	237	235	212
21	O,P DDD	237	235	178
22	O,P DDE	318	226	–
21	Pendimethalin	281	252	–
23	Endosulphan Sulphate	420	387	272
24	Dialdrin	277	263	261
25	Heptachlor	272	237	266
25	Cypermethrin	181	263	–
27	Deltamethrin	253	181	–
28	Butachlor	176	160	–

2.7 *Uncertainty Calculation*

Calculation of the total uncertainty depends on four major uncertainty i.e. uncertainty due to weighing, purity of the standard and repeatability. The expanded uncertainty of the pesticides are twice that of total uncertainty. Expanded uncertainty values lies in three ranges viz., (a) 15–20% (b) 11–15% and (c) $\leq 10\%$ (Table 4). In spinach, all 28 pesticides expanded uncertainty lies below $\leq 10\%$. So, the method followed for all pesticides is reliable in analysing pesticides from the spinach matrices taken. In this study, there are three major sources of combined uncertainty i.e. repeatability of analytes in spiked samples, preparation of the calibration standard solutions (weighing, diluting) and purity of standards.

Table 3 Shows recovery, mean recovery, standard deviation (S.D), relative standard deviation (RSD) of organochlorine, synthetic pyrethroids and herbicides pesticides from spiked spinach matrix at 0.05 ppm

S. No.	Pesticide	RT	Spiking conc (PPM)	R1		R2		R3		Mean Rec. percent	Mean Rec. amount	S.D.	R.S.D
				Amount recovered	Rec percent	Amount recovered	Rec percent	Amount recovered	Rec percent				
1	Alpha-HCH	9.80	0.05	0.042	84	0.046	92	0.046	92	89.33	0.04	0.002	5.170
2	Dicofol	10.80	0.05	0.046	92	0.043	86	0.048	96	91.33	0.05	0.003	5.511
3	Beta-HCH	11.25	0.05	0.046	92	0.042	84	0.045	90	88.67	0.04	0.002	4.695
4	Gamma HCH	11.57	0.05	0.036	90	0.039	99	0.039	96	95	0.04	0.002	4.558
5	Delta HCH	12.86	0.05	0.049	98	0.039	78	0.043	86	87.33	0.04	0.005	11.526
6	Heptachlor	15.78	0.05	0.04	80	0.043	86	0.046	92	83.33	0.04	0.003	6.977
7	Alachlor	15.87	0.05	0.045	90	0.041	82	0.039	78	83.33	0.04	0.003	7.332
89	Aldrin	18.05	0.05	0.043	86	0.043	86	0.04	80	84	0.04	0.002	4.124
10	Pendimethalin	21.06	0.05	0.042	84	0.046	92	0.047	94	90	0.05	0.003	5.879
11	O,P DDE	23.21	0.05	0.047	94	0.042	84	0.043	86	88	0.04	0.003	6.013
12	Alpha-Endosulphan	23.52	0.05	0.044	88	0.039	78	0.048	96	87.33	0.04	0.005	10.327
13	Butachlor	24.22	0.05	0.041	82	0.048	96	0.04	80	86	0.04	0.004	10.137
14	Dialdrin	25.53	0.05	0.046	92	0.048	96	0.048	96	94.67	0.05	0.001	2.440
15	P,P DDE	25.68	0.05	0.039	78	0.039	78	0.046	92	82.67	0.04	0.004	9.778
16	O,P DDD	26.30	0.05	0.046	92	0.04	80	0.045	90	87.33	0.04	0.003	7.362
17	P,P DDT	26.40	0.05	0.045	90	0.043	86	0.048	96	90.67	0.05	0.003	5.551
18	Beta-Endosulphan	28.14	0.05	0.047	94	0.041	82	0.044	88	88	0.04	0.003	6.818
19	P,P DDD	29.47	0.05	0.035	70	0.039	78	0.039	78	75.33	0.04	0.002	6.131
20	O,P DDT	29.72	0.05	0.038	76	0.04	80	0.041	82	79.33	0.04	0.002	3.851
21	Endosulphan Sulphate	32.60	0.05	0.043	86	0.045	90	0.044	88	88	0.04	0.001	2.273
22	Bifenthrin	41.72	0.05	0.048	96	0.04	80	0.049	98	91.33	0.05	0.005	10.802

(continued)

Table 3 (continued)

S. No.	Pesticide	RT	Spiking conc (PPM)	R1		R2		R3		Mean Rec. percent	Mean Rec. amount	S.D.	R.S.D
				Amount recovered	Rec percent	Amount recovered	Rec percent	Amount recovered	Rec percent				
23	Fenpropathrin	42.19	0.05	0.045	90	0.04	80	0.042	84	84.67	0.04	0.003	5.945
24	Lambda Cyhalothrin	47.44	0.05	0.037	74	0.04	80	0.043	86	80	0.04	0.003	7.500
25	Beta Cyfluthrin	52.7–52.9	0.05	0.039	78	0.045	90	0.045	90	86	0.04	0.003	8.056
26	Cypermethrin	53.03–53.44	0.05	0.048	96	0.04	80	0.043	86	87.33	0.04	0.004	9.255
27	Fenvalarate	56.23	0.05	0.047	94	0.043	86	0.042	84	88	0.04	0.003	6.013
28	Fluvalinate	56.9–57.2	0.05	0.043	86	0.041	82	0.041	82	83.33	0.04	0.001	2.771
29	Deltamethrin	58.63	0.05	0.048	96	0.042	84	0.043	86	88.67	0.04	0.003	7.251

Table 4 Results of individual and combined uncertainties with expanded uncertainty for of organochlorine, synthetic pyrithroids and herbicides pesticides from spiked spinach matrix at 0.05 ppm

S. No.	Pesticide	Retention Time	Purity (%)	Wt. std	Un-certainty	SU1	U1	U2	SD-recovery	Mean-recovery	Replication	U3	U	2U
1	Alpha-HCH	9.80	99.6	1.24	0.005	0.003	0.003	4.03226E-05	0.002	0.04	3	0.0289	0.0012	0.0023
2	Dicofol	10.80	99.5	1.91	0.005	0.003	0.003	2.61780E-05	0.003	0.05	3	0.0346	0.0017	0.0035
3	Beta-HCH	11.25	98.8	1.37	0.005	0.003	0.003	3.64964E-05	0.002	0.04	3	0.0289	0.0012	0.0023
4	Gamma HCH	11.57	99.6	1.86	0.005	0.003	0.003	2.68817E-05	0.002	0.04	3	0.0289	0.0012	0.0023
5	Delta HCH	12.86	99.7	1.48	0.005	0.003	0.003	3.37838E-05	0.005	0.04	3	0.0722	0.0029	0.0058
6	Heptachlor	15.78	99.3	1.27	0.005	0.003	0.003	3.93701E-05	0.003	0.04	3	0.0433	0.0017	0.0035
7	Alachlor	15.87	99.8	1.25	0.005	0.003	0.003	4.00000E-05	0.003	0.04	3	0.0433	0.0017	0.0035
8	Aldrin	18.05	99.2	1.23	0.005	0.003	0.003	4.06504E-05	0.002	0.04	3	0.0289	0.0012	0.0023
9	Pendimethalin	21.06	98.9	1.73	0.005	0.003	0.003	2.89017E-05	0.003	0.05	3	0.0346	0.0017	0.0035
10	O,P DDE	23.21	97.2	1.8	0.005	0.003	0.003	2.77778E-05	0.003	0.04	3	0.0433	0.0017	0.0035
11	Alpha-Endosulphan	23.52	99.5	1.45	0.005	0.003	0.003	3.44828E-05	0.005	0.04	3	0.0722	0.0029	0.0058
12	Butachlor	24.22	99.5	1.56	0.005	0.003	0.003	3.20513E-05	0.004	0.04	3	0.0577	0.0023	0.0046
13	Dialdrin	25.53	99	1.56	0.005	0.003	0.003	3.20513E-05	0.001	0.05	3	0.0115	0.0006	0.0012
14	P,P DDE	25.68	99.4	1.84	0.005	0.003	0.003	2.71739E-05	0.004	0.04	3	0.0577	0.0023	0.0046
15	O,P DDD	26.30	99.7	1.87	0.005	0.003	0.003	2.67380E-05	0.003	0.04	3	0.0433	0.0017	0.0035
16	P,P DDT	26.40	96	1.82	0.005	0.003	0.003	2.74725E-05	0.003	0.05	3	0.0346	0.0017	0.0035
17	Beta-Endosulphan	28.14	99.5	1.57	0.005	0.003	0.003	3.18471E-05	0.003	0.04	3	0.0433	0.0017	0.0035
18	P,P DDD	29.47	96	1.46	0.005	0.003	0.003	3.42466E-05	0.002	0.04	3	0.0289	0.0012	0.0023
19	O,P DDT	29.72	99.6	1.83	0.005	0.003	0.003	2.73224E-05	0.002	0.04	3	0.0289	0.0012	0.0023
20	Endosulphan Sulphate	32.60	99	1.74	0.005	0.003	0.003	2.87356E-05	0.001	0.04	3	0.0144	0.0006	0.0012
21	Bifenthrin	41.72	99.5	1.46	0.005	0.003	0.003	3.42466E-05	0.005	0.05	3	0.0577	0.0029	0.0058
22	Fenpropathrin	42.19	99.5	2.1	0.005	0.003	0.003	2.38095E-05	0.003	0.04	3	0.0433	0.0017	0.0035
23	Lambda Cyhalothrin	47.44	98.5	1.54	0.005	0.003	0.003	3.24675E-05	0.003	0.04	3	0.0433	0.0017	0.0035

(continued)

Table 4 (continued)

S. No.	Pesticide	Retention Time	Purity (%)	Wt. std	Un-certainty	SU1	U1	U2	SD-recovery	Mean-recovery	Replication	U3	U	2U
24	Beta Cyfluthrin	52.7–52.9	99.5	1.36	0.005	0.003	0.003	3.67647E-05	0.003	0.04	3	0.0433	0.0017	0.0035
25	Cypermethrin	53.03–53.44	99.7	1.34	0.005	0.003	0.003	3.73134E-05	0.004	0.04	3	0.0577	0.0023	0.0046
26	Fenvalarate	56.23	99.3	1.89	0.005	0.003	0.003	2.64550E-05	0.003	0.04	3	0.0433	0.0017	0.0035
27	Fluvalinate	56.9–57.2	98.8	1.87	0.005	0.003	0.003	2.67380E-05	0.001	0.04	3	0.0144	0.0006	0.0012
28	Deltamethrin	58.63	99.5	1.45	0.005	0.003	0.003	3.44828E-05	0.003	0.04	3	0.0433	0.0017	0.0035

SU1 Standard uncertainty of analytical standards
U1 Relative standard uncertainty of analytical standards
U2 Relative standard uncertainty of weighing
U3 Uncertainty associated with precision
U Combined uncertainty
2U Expanded uncertainty

3 Material and Methods

3.1 Reagents and Materials

Reference standards of pesticides taken for study were purchased from Sigma-Aldrich (Germany). Ethyl Acetate (EtAc), n-Hexane grade, Sodium Sulphate Anhydrous, Magnesium sulfate (anhy. $MgSO_4$) were purchased from Merck (Germany), Primary Secondary Amine (PSA) and graphitized carbon black sorbent were obtained from Varian (Varian Incorporated, Harbor City, CA, USA), and Supelco (Bellefonte, PA, USA), respectively.

3.2 Sampling

Sample of spinach was purchased from local vegetable mandi of Satna District, M. P. (India). Purchased sample was transported in ice pack to the laboratory in cool-box, stored at 4 °C and extracted within 48 h.

3.3 Sample Preparation

QuEChERS (quick, easy, cheap, effective, rugged and safe) method with some modification were followed for sample preparation [1]. Spinach was homogenized in a mixer grinder. Fluoroethylenepropylene (FEP) centrifugation tube (Tarson make) 50 ml was taken for weighing fifteen gram of thoroughly homogenized sample and 30 ml of Ethyl acetate was added and shaken for 1 min Vortex shaker. Ten gram anhydrous Na_2SO_4 was added and shaken for 1 min. after the tubes were centrifuged at 5,000 rpm on the centrifuge machine at about 5 min. Highly pigmented foods, such as spinach, required cleanup prior to analysis for pesticide residues [6]. 6.0 ml of extract upper layer was transferred into a 15 ml centrifuge tube, and 0.9 g anhydrous $MgSO_4$, 0.25 g PSA and 0.25 g activated charcoal were added and shaken for 10 min by vortex shaker (Torson make) at 50 rpm, later these tubes were centrifuged at 5,000 rpm for 5 min. The supernatant 4 ml was dried and finally made up to 1 ml for injection in GC-ECD and GC-MS.

3.4 Stock Standard Solution

Certified Reference Materials (CRM) having specific purity of each pesticide were taken for the study. The stock solution of each pesticide was prepared. Clean and dried 10 ml standard volumetric was taken for weighing CRM of individual pesticide on analytical balance pan (Mettler, Toledo) maximum up to 4 mg. Few drops of acetone was added to dissolve the CRM and further hexane was added to fill standard volumetric flask up to the mark. Prepared standard stock solution and working standards were stored in deep freezer at −20 °C for further analysis.

3.5 *Instrument Operating Parameters*

Validation of method, uncertainty calculation and residue screening of spinach samples were conducted using two instruments. First samples were analysed by GC-ECD for the initial screening of the samples for presence of pesticides. Secondly concentrated samples were analysed by GC-MS in full scan mode for further confirmation. By Retention Time (RT) and Mass Spectrum (MS) match, the presence of pesticides in samples was confirmed. Final quantification was carried out on GC-ECD.

3.5.1 GC-ECD

GC-ECD Agilent 7890B (7693 auto sampler) with DB-5MS fused silica capillary column (Agilent J&W GC column, 5% Phenylated methyl siloxane, 30 m length 0.25 mm i.d. × 0.25 µm film thickness) attached was used for screening and quantification of pesticide.

Column:	DB-5 (length: 30 m, I.D.: 0.25 mm, film thickness: 0.25 µm)
Oven temperature program:	170 °C (2 min), @ 10 °C, 280 °C (10 min)
Detector temperature:	300 °C
Injector temperature:	250 °C
Carrier gas (He) flow rate:	1.35 ml/min
Injection volume for standard and sample:	1 µl

3.5.2 GC-MS

Agilent Technologies GC Model 7890A with mass (5975C inert XL EI/CI MSD) detector was used for pesticide residue analysis. Injector port temperature was set at 280 °C. Agilent DB-5MS fused silica capillary GC column (5% Phenylated methyl siloxane, 30 m length × 0.25 mm i.d. × 0.25 µm film thickness) was used with linear flow at 1 ml/min. 1 micro litre of concentrated extract was injected in splitless mode with carrier gas helium (grade-1). The pesticides were separated with 40.75 min oven programming starts with initial temperature 50 °C for 2 min with a ramp rate of 8 °C/min up to a temperature of 280 °C with a hold time of 10 min. The injector was operated in splitless mode at 280 °C temperature. The interface, ion source and quadruple temperatures were set at 280 and 230 and 150 °C respectively. The mass spectrometer was operated in electron impact (70 eV), selected ion monitoring (SIM) and Scan mode with solvent delay time 3 min.

3.6 *Method Development and Validation*

Before any analytical method to be used routinely in the laboratory, it must be validated. Validation is the test for the suitability of a particular method for collecting analytical data. There are several key parameters for ensuring a robust method, they are—Recovery, Selectivity (Specificity), Calibration, Repeatability, Reproducibility, Limit of Detection (LOD), Limit of quantitation (LOQ) [7], SANCO/2007/3131 (Supersedes Document No. SANCO/10232/2006).

3.7 *Uncertainty Calculation*

Combined uncertainty (U) was calculated at the level of 0.05 ppm as per the statistical procedure [8–10] Uncertainty was calculated from the 3 sources namely:

(1) Relative Standard Uncertainty of Analytical Standards (U1):

Some of the CRMs are not mentioned with uncertainty value by their manufacturers. To calculate standard uncertainty (SU1), normal distribution is assumed in the below equation,

$$\mathrm{SU1} = (\mathrm{U_c}/2);$$

where,

U_c is the uncertainty value of the CRM purity percentage mentioned in the certificate,

Standard uncertainty (SU1) was used to calculate the Relative standard uncertainty (U1) by the formula:

$$\mathrm{U1} = (\mathrm{SU1} \times 100)/\text{percent purity}$$

(2) Relative Standard Uncertainty in Weighing (U2):

Assuming normal distribution for uncertainty caused by weighing is calculated by formula:

$$U2 = (U_b/2)/w_i$$

where,

U_b is the uncertainty of weighing balance at 95% confidence level which is 0.0001, w_i is the weight measured (mg) of individual pesticide

(3) Relative standard uncertainty associated with precision (U3):

Uncertainty associated to the precision or repeatability of results are due to the random errors in the sample processing and instrumental analysis. To measure U3 three replicate recoveries are calculated for its Mean (m), Standard deviation (SD)

$$U3 = s/(\sqrt{}n \times x)$$

where: s = standard deviation comes from the recovery study, n = number of replications and x = mean value of the concentration recovered.

The combined uncertainty (U) is calculated by Eq. 5.

$$U = x[(U1)2 + (U2)2 + (U3)2]^{1/2}$$

Expanded uncertainty (2U) is twice of combined uncertainty (U) at 95% confidence level.

Uncertainties values of each pesticide is shown in Table 3.

4 Conclusion

For routine pesticide residue analysis, in-house method was validated. The study focuses to overview analytical performance of accepted sample preparation method combined with analytical instrumentation. Performance of the method indicates that the majority of target compounds comply with current regulatory requirements. To improve confidence in the measurements, independent, external quality control materials are additionally applied. For study, pesticides of different class, high persistence and widely used in India, were taken. In some cases, measured values fell outside of the targeted range, due to specific properties, strong matrix effect. For those compounds method performance parameters could not be established (in the relevant matrix) and only individually optimized sample preparation (additional or special clean-up) and instrumental methods have to be applied. Results of the method applied in the study were repeatable, reproducible and gives good recovery with uncertainty range within permissible limit. The results obtained indicate that the developed method is rapid, accurate, selective, and reproducible. All pesticides show linear behaviour for the conc. range of 1.0–0.003 ppm standard mixture analyzed by GC-ECD and correlation coefficient (R2) was found to be 0.993 for lambda cyhalothrin. The method has been successfully applied for the analysis of vegetables samples. It can be used for the routine analysis of multi residue in different vegetables matrix. Validation of this method meet all requirements as per standard guidelines.

Acknowledgements The authors are highly grateful to the Director, Institute of Pesticide Formulation Technology for allowing them to use the research facilities for the work.

Conflict of Interest The authors declare no conflict of interest.

References

1. Anastassiades M, Lehotay SJ, Stajnbaher D, Schenck FJ (2003) Fast and easy multiresidue method employing acetonitrile extraction/partitioning and "dispersive solid-phase extraction" for the determination of pesticide residues in produce. J AOAC Int 86:412–431
2. Hernandez-Borges J, Cabrera JC, Rodriguez-Delgado MA et al (2009) Analysis of pesticide residues in bananas harvested in the Canary Islands (Spain). Food Chem 113:313–319
3. Lehotay SJ, De Kok A, Hiemstra M et al (2005) Validation of a fast and easy method for the determination of residues from 229 pesticides in fruits and vegetables using gas and liquid chromatography and mass spectrometric detection. J AOAC Int 88:595–614
4. Nguyen TD, Yu JE, Lee DM, Lee GH (2008) A multiresidue method for the determination of 107 pesticides in cabbage and radish using QuEChERS sample preparation method and gas chromatography mass spectrometry. Food Chem 110:207–213
5. Paya P, Anastassiades M, Mack D et al (2007) Analysis of pesticide residues using the Quick Easy Cheap Effective Rugged and Safe (QuEChERS) pesticide multiresidue method in combination with gas and liquid chromatography and tandem mass spectrometric detection. Anal Bioanal Chem 389:1697–1714
6. Lehotay SJ (2007) Determination of pesticide residues in foods by acetonitrile extraction and partitioning with magnesium sulfate: collaborative study. J AOAC Int 90:485–520
7. EURACHEM/CITAC Guide CG 4, EURACHEM/ CITAC Guide (2000) Quantifying uncertainty in analytical measurement, 2nd ed
8. Sanyal D, Rani A, Alam S et al (2011) Development validation and uncertainty measurement of multi-residue analysis of organochlorine and organophosphorus pesticides using pressurized liquid extraction and dispersive-SPE techniques. Environ Monit Assess 182:97–113
9. Saini MK, Mishra S, Alam S et al (2016) Method development and validation of multiclass pesticide residues and metabolites in wheat by GC-ECD and GC-MS. Asian J Res Chem 9 (1):1–9
10. Mishra S, Saini MK, Alam S et al (2015) Determination of 25 pesticides By GC-ECD And MSD with measurement of uncertainty in tomato using modified QuEChERS technique. J Appl Chem 4(1):221–231
11. Corley J (2002) Best practices in establishing detection and quantification limits for pesticide residues in foods. Rutgers, The State University of New Jersey, North Brunswick, NJ, USA
12. Corley J (2003) Best practices in establishing detection and quantification limits for pesticide residues in foods. In: Handbook of residue analytical methods for agrochemicals. Wiley, New York, USA, pp 59–74
13. Corley J (2005) Handbook of residue analytical methods for agrochemicals. Wiley, p. 59

Evaluation of Formation Constants of Mixed-Ligands Complexes of Cadmium(II) with Some Amino Acids (Alanine, Serine) and 4,4,4-Trifluoro-1-(2-Napthyl)Butane-1,3-Dione at DME in 60% Acetonitrile Medium

Suman Meena and Rajni Grover

Abstract Intensive polarographic study of Cd(II) with amino acids (L-Serine, DL-Alanine) and the ligand 4,4,4-trifluoro-1-(2-napthyl)butane-1,3-dione have been carried out keeping constant ionic strength ($\mu = 1$), at 308 K, by using KNO_3 in 60% Acetonitrile media. It was found that the reduction of all the systems taken, are diffusion controlled and reversible involving two electrons. For the simple system of Cd(II) with 4,4,4-trifluoro-1-(2-napthyl)butane-1,3-dione and Cd(II) with amino acids, the stability constants were determined first by DeFord and Hume and then evaluated by Schaap and McMaster method.

Keywords Amino acids · Metal complexes · Reduction polarographic reversible Mixed-ligands complexes

1 Introduction

Metal ion, in solution, with two or more different ligands form mixed-ligand complexes. In biological active substance, mixed chelation generally takes place as millions of potential ligands are about to compete for metal ions, thus these mixed-ligands complexes play an important role in many of the biological processes.

Amino acids shows great significance in pharmaceutical and biological fields and are well known for their strong tendency to form complexes with transition metal ions, also they are directly involved in all the enzymatic metabolic processes. The amino acids—metal ion complexes are used in pharmacy, cancer therapy, and industry [1]. Amino acids, because of having good chelating ability with transition metal ions, part an important role in pharmacy and biology [2, 3].

S. Meena (✉) · R. Grover
Rajdhani College, Delhi University, New Delhi 110015, India
e-mail: sumanjay21@gmail.com

V. S. Parmar et al. (eds.), *Green Chemistry in Environmental Sustainability and Chemical Education*, https://doi.org/10.1007/978-981-10-8390-7_18

Seri ne and alanine are used as strong ligands in the study where serine [4] is one of the naturally occurring proteinozenic amino acids. It is of great importance in metabolism as it participates in the pyrimidines and purine biosynthesis. It is the precursor to several amino acids including tryptophane, cysteine and glycine. It is essential for immune system, metabolism of fat, and tissue growth as it assists in fabrication of antibodies and immunoglobuins. On the other hand, DL-Alanine amino acid helps the body to convert the glucose to energy and causes the elimination of surplus toxins from the liver. Alanine helps in protecting cells from being damaged, during extreme aerobic activity [5], when the body cannibalizes muscles protein to help produce energy. It's help regulate blood sugar and adrenal gland function.

Most of the Cd available is used in Ni-Cd batteries and the remaining is used mainly for pigments coating and plating and as stabilizers for plastics [6, 7]. Cadmium has the ability to absorb neutrons so it is used as a barrier to control nuclear fission. Cadmium is amongst the most toxic of the heavy metals. It primarily affects the cardiovascular system, the kidney, and is related to diabetes, cancer, and heart diseases. Many authors calculated the stability constants of Pb(II) and Cd(II) ion with some of the amino acids and some bicarboxylic acids at DME at the requisite temperature [8–10]. Formation constant of Cd(II) with penicillamine and Histidine have been studied by Jadwiga et al. [11]. Paliwal and coworkers [12–14], studied the stability constants of Pb(II) ion complexes with some amino acids and β-picoline and Cd(II) ion complexes with some amino acids and pyridine.

Kalpana et al. [15] have studied complexes of Pb(II) and Cd(II) ions with thioglycolate-Tartarate systems in aqueous-methanol media polarographically. Khan and Jain [16] have reported the formation of mixed-ligands complexes of Cd(II) with L-Amino acids as primary ligand and nicotinic acid as secondary ligand. A huge number of ternary complexes of Cd(II) have been studied by polarographic technique during last years [17–20]. Paliwal et al. [21, 22] have studied the mixed-ligands complexes of Cd(II) ions with pyridine, lysine, methionine, β-picoline and some hydroxy acids. Many other workers [23–26] have also studied the mixed-ligands complexes of Cd(II) ions with various amino acids and carboxylic acids. Joshi et al. [27] have studied mixed-ligands complexes of Cu(II) with Ethylenediamine, 1,2-Diaminopropane, 1,3-diaminopropane, and 2,2-Bipyridylamine potentiometrically.

The literature study reveals that there is insufficient data present on the 4,4,4-trifluoro-1-(2-napthyl)butane-1,3-dione, amino acids mixed-ligands complexes of Cd(II). On the other hand the 4,4,4-trifluoro-1-(2-napthyl)butane-1,3-dione which has been used in present paper as a weak ligand is a biochemical reagent, chelator and ion probes for spectroscopy and fluorescent indicator. The present paper deals with the determination of the stability constants of mixed-ligands complexes of Cd(II) with 4,4,4-trifluoro-1-(2-napthyl)butane-1,3-dione and some amino acids.

2 Experimental

2.1 *Reagents*

All reagent grade chemicals were used for preparation of solution containing 0.1 M Cd(II) with various concentration of ligands. Sodium salts of Alanine and Serine were used as complexing agents and as weak ligands 4,4,-trifluoro-1-(2-napthyl) butane 1,3-dione was used which is soluble in organic solvents only. 60% acetonitrile was used as media for preparation of solutions for the experiment. To maintain constant ionic strength of the solution, supporting electrolyte KNO_3 was used and the role of Triton X-100 was as maximum suppressor.

2.2 *Apparatus*

To record polarograms of the complexes, a CL-**362 polarographic analyser** using saturated calomel electrode as reference electrode was used where the microelectrode used is dropping mercury electrode (D.M.E.). Haake-type thermostat was used to maintain the solution under test at constant temperature. The pH of the test solutions were adjusted with the help of digital pH meter. The capillary had the following characteristic m = 4.62 mg/s, t = 3 s and height of mercury column h = 43 cm. Removal of dissolved oxygen for polarographic estimation was done by passing purified nitrogen gas through the test solution for 15 min. To obtain the required polarogram, the variation of current with reference to potential was recorded. In all the test solutions, the variation in the concentration of the stronger ligand was taken keeping that of the weaker ligand as constant.

3 Results and Discussion

The simple complexes of Cd(II) with DL-Alanine, Serine and 4,4,4-triluoro-1-(2-napthyl)butane-1,3-dione were investigated and the evaluation of their overall formation constants were done by DeFord and Hume's [24] method from the cathodic shift in half-wave potential as a function of ligand concentration. The reduction of Cd(II) was reversible with the involvement of two electrons and entirely diffusion controlled in all the systems. The conclusion have been drawn from the slope of the plots of log $i/i_d - i$ versus $E_{d.e.}$ which were straight line.

The formation constants of Cd(II) simple complexes with various ligands systems are summarized in Table 1.

Table 1 Stability constants of simple Cd(II) complexes with various ligands in aqueous-non aqueous medium at 308 K

S. No.	Metal	Ligand	Overall formation constant		
			$\log\beta_1$	$\log\beta_2$	$\log\beta_3$
1.	Cd(II)	4,4,4-trifluoro-1-(2-napthy)butane-1,3-dione	1.7781	3.47	5.38
2.	Cd(II)	DL-Alanine	3.27	5.079	8.27
3.	Cd(II)	Serine	2.69	5.07	8.14

3.1 The Mixed-Ligands Complexes

The maximum possible co-ordination number of the transition metal Cd(II) ion is six. [Cd(Maybridge)], [Cd(amino acid)] and [Cd(amino acid)(Maybridge)$_2$] complexes would be predicted binding with the two dissimilar bidentate ligands. The concentration values of stronger ligands (Alanine/Serine) varied in the range 0.001–0.007 M and that of the weaker ligands (Maybridge) are kept constant at two concentration values (0.075 and 0.15 M).

The polarograms were plotted by observing current-voltage readings. From the slopes of log plots of $E_{d.e.}$ versus log $i/i_d - i$ this system reduced reversibly at d.m. e. at 308 K. A cathodic shift in half-wave potential is observed as a function of DL-Alanine and serine concentrations. The magnitude of the shift in the half-wave potential in presence of weak ligand (Maybridge) is greater than obtained for the simple Cd(II)-DL-Alanine and serine systems. It indicates the formation of mixed-ligands complex with the increase in concentration of DL-Alanine and serine and it was observed that half-wave potential ($E_{1/2}$) shifted towards more negative direction. The extended Schaap and McMaster's [25] method was applied and Leden's [26] graphical extrapolation method to evaluate the A, B, C, D which have been recorded in Tables 2 and 3 for both fixed concentrations of weak ligand.

From the expressions for B and C values, the values of log β_{11}, $\log\beta_{12}$ and $\log\beta_{21}$ were calculated from the two equations given below.

$$\text{“}B = \beta_{10} + \beta_{11}[Y] + \beta_{12}[Y]^2\text{”} \quad (1)$$

$$\text{“}C = \beta_{20} + \beta_{21}[Y]\text{”} \quad (2)$$

Table 2 The values of A, B, C and D for mixed-ligands complex of Cd(II)-4,4,4-triluoro-1-(2-napthyl)butane-1,3-dione and DL-Alanine system at 308 K

4,4,4-triluoro-1-(2-napthyl)butane-1,3-dione Maybridge mole/litre (fixed)	log A	log B	log C	log D
0.075 C_x	0.8750	4.3010	6.6020	9.2585
0.15 C_x	1.3979	4.4771	7.27	9.8450

Table 3 The values of A, B, C and D for mixed-ligands complex of Cd(II)-4,4,4-triluoro-1-(2-napthyl)butane-1,3-dione and Serine system at 308 K

Maybridge mole/litre (fixed)	log A	log B	log C	log D
0.075	0.4771	3.7781	6.9540	8.9030
0.15	0.6020	3.9541	7.1139	8.9542

Here Y indicates concentration of weaker ligand (Maybridge).

The results are summarised in the form of Schemes 1 and 2 and formation constants of mixed-ligands complex systems are recorded in Table 4.

According to the theoretical principle, $\log_{30}$ should be equal to log D when all the coordinated maybridge are replaced by DL-Alanine and serine. A little higher value of log D than that of log β_{30} may be due to the fact that while observing $\beta_{30,}$ there is only DL-Alanine as complexing agent present in solution while for D, Maybridge is also present in the solution. The presence of maybridge in solution probably makes the value of D higher than that of β_{30}. This observation shows that the stability of a complex species is affected by the presence of a foreign substance in solution.

The results show that the mixed-ligands complex may be attributed to a significant difference in the sizes and charge neutralizing capacities of two ligands. The complex [Cd(X)(Y)] able to add X more simply than to add Y showing that X is stronger ligand and are more charge neutralizing. The mixed-ligands complex $[Cd(X)(Y)_2]^{-1}$ has tendency to substitute and not to add Y, indicating that X has greater complexing tendency than Y which shows mixed-ligand complex formation is favoured. The tendencies of "$[Cd(X)_2]$" to add X and Y favor mixed-ligands complex formation like wise. The tendencies of $[Cd(X)_2]$ and "$[Cd(Y)_2]$" to add Y

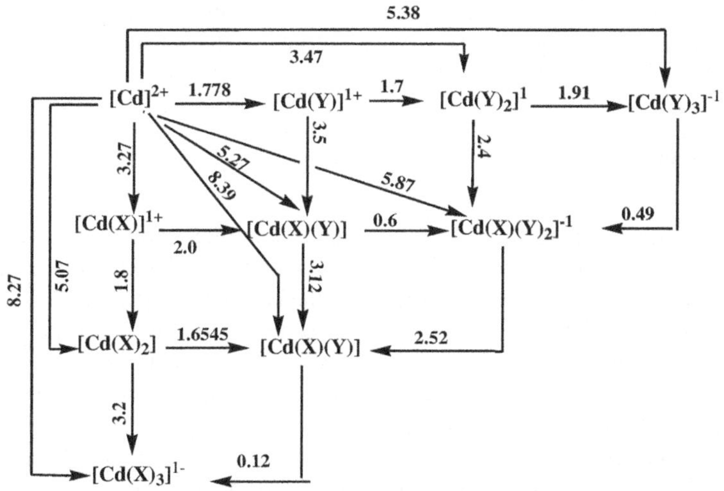

Scheme 1 Cd(II)-4,4,4-trifluoro-1-(2-napthyl)butane-1,3-dione DL-Alanine system at 308 K

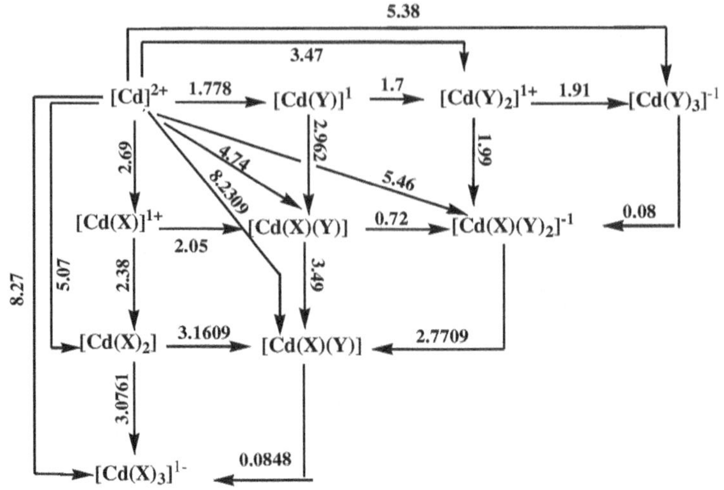

Scheme 2 Cd(II)-4,4,4-trifluoro-1-(2-napthyl)butane-1,3-dione-L-Serine system at 308 K

Table 4 Stability constant of mixed-ligands of complexes Cd(II) amino acid-Maybridge system at 308 K

Maybridge-DL-Alanine system	Maybridge-Serine system
$\log\beta_{11} = 5.27$	$\log\beta_{11} = 4.74$
$\log\beta_{12} = 5.87$	$\log\beta_{12} = 5.46$
$\log\beta_{21} = 8.39$	$\log\beta_{21} = 8.23$

and X, respectively show the easier approach of X to $[Cd(Y)_2]$ to form $[Cd(X)(Y)_2]^{-1}$ than the approach of Y to $[Cd(X)_2]$ to form $[Cd(X)_2Y]^{-1}$.

The maximum stoichiometric ratio is 1:3 of each of the ligand with Cd(II) shows that the chelation takes place. The representation of all the complexes of present system and equilibria amongst them are as shown in the schemes.

4 Discussion

Schemes 1 and 2 explained the mixed-ligands complex formation. Electrostatic effects and entropy must be correlated to the main part of the difference in log K, therefore charge complex species formed. To the complexes "$[Cd(X)]^+$ (X = amino acids) and $[Cd(Y)]^+$ (Y = Maybridge)", the tendency to add X can be compared easily. The stability constants log values of the above mentioned complexes are (1.8, 3.5), (2.38, 2.96) at 308 K for Cd(II)-DL-Alanine-Maybridge and Cd(II)-Serine-Maybridge system, respectively.

To the complexes "$[Cd(X)]^{+1}$ and $[Cd(Y)]^{+1}$", the tendency to add Y may also be compared. For Cd(II)-Maybridge-Alanine and Cd(II)-Maybridge-Serine system, the

Table 5 $F_{io}[(X)]$ function values for Cd(II)-DL-Alanine-Maybridge system at 308 K, ionic strength (μ) = 1.0 (KNO_3), Triton X-100 = 0.0002% [Cd(II)] = 1.0 mM, Maybridge acid = 0.075 M (fixed)

C_X (moles/litre)	i_d (μA)	$E_{1/2}$ (−V vs. S.C.E)	$F_{00}[(X)]$	$F_{10}[(X)] \times 10^4$	$F_{20}[(X)] \times 10^6$	$F_{30}[(X)] \times 10^9$
0.000	8.40	0.650	–	–	–	–
0.001	8.02	0.693	27.82	2.682	6.82	2.8
0.002	7.83	0.705	71.00	3.50	7.5	1.75
0.003	7.23	0.714	144.6	4.84	9.49	1.830
0.004	7.08	0.721	261.93	6.52	11.30	1.825
0.005	6.98	0.727	416.51	8.3102	13.82	1.964
0.006	6.92	0.733	654.34	10.889	14.81	1.801
0.007	6.82	0.737	910.79	12.99	17.13	1.85

log A = 0.8750, log B = 4.3010, log C = 6.6020, log D = 9.2585, C_x = concentration of DL-Alanine in moles $litre^{-1}$

Table 6 $F_{io}[(X)]$ function values for Cd(II)-DL-Alanine-Maybridge system at 308 K, ionic strength (μ) = 1.0 (KNO_3), Triton X-100 = 0.0002% [Cd(II)] = 1.0 mM, Maybridge = 0.15 M (fixed)

C_X (moles/litre)	i_d (μA)	$E_{1/2}$ (−V vs. S.C.E)	$F_{00}[(X)] \times 10^2$	$F_{10}[(X)] \times 10^4$	$F_{20}[(X)] \times 10^7$	$F_{30}[(X)] \times 10^{10}$
0.000	8.40	0.650	–	–	–	–
0.001	7.86	0.700	1.65	6.4013	3.4	1.500
0.002	7.48	0.719	2.010	10.0013	3.50	1.050
0.003	7.29	0.729	5.689	15.6314	4.20	0.766
0.004	7.00	0.737	10.17	22.9314	4.982	0.0770
0.005	6.90	0.745	16.67	31.3550	5.67	0.754
0.006	6.74	0.751	26.41	42.3610	6.56	0.776
0.007	6.52	0.755	38.15	53.0819	7.1545	0.750

log A = 1.3979, log B = 4.4771, log C = 7.27, log D = 9.8450, C_x = concentration of DL-Alanine in moles $litre^{-1}$

log K values are (2.0, 1.7), (2.05, 1.7) respectively at 308 K. This confirms the addition of Maybridge is preferred to $[Cd(X)]^{+1}$ as compared to $[Cd(Y)]^{+1}$.

For the addition of X to [Cd(X)(Y)] and $[Cd(Y)_2]$, the log values are (3.12, 2.4), (3.49, 1.99) for Cd(II)-Maybridge-Alanine, Cd(II)-Maybridge-Serine system, respectively. This indicates that the mixed-ligands complexation is favoured.

For the addition of Y to [Cd(X)(Y)] and $[Cd(X)_2]$, the log K values are (0.6, 1.65), (0.72, 3.16) for Cd(II)-Maybridge-Alanine, Cd(II)-Maybridge-Serine system, respectively.

Table 7 $F_{i0}[(X)]$ function values for Cd(II)-Serine-Maybridge system at 308 K, ionic strength (μ) = 1.0 (KNO_3), Triton X-100 = 0.0002% [Cd(II)] = 1.0 mM, Maybridge = 0.075 M (fixed)

C_X (moles/litre)	i_d (μA)	$E_{1/2}$ (−V vs. S.C.E)	$F_{00}[(X)]$	$F_{10}[(X)] \times 10^3$	$F_{20}[(X)] \times 10^6$	$F_{30}[(X)] \times 10^8$
0.000	8.40	0.650	–	–	–	–
0.001	7.80	0.685	16.805	15.805	9.805	8.05
0.002	7.57	0.701	53.812	26.405	10.23	6.15
0.003	7.36	0.712	126.28	41.76	11.92	9.73
0.004	7.12	0.719	226.44	56.36	12.59	8.97
0.005	6.91	0.725	370.00	73.8	13.56	9.12
0.006	6.81	0.731	554.68	92.28	14.38	8.97
0.007	6.69	0.735	780.46	111.35	15.050	8.64

log A = 1.477, log B = 3.7781, log C = 6.95, log D = 8.9030, C_x = concentration of Serine in moles $litre^{-1}$

Table 8 $F_{i0}[(X)]$ function values for Cd(II)-Serine-Maybridge system at 308 K, Ionic strength (μ) = 1.0 (KNO_3), Triton X-100 = 0.0002% [Cd(II)] = 1.0 mM, Maybridge = 0.15 M (fixed)

C_X (moles/litre)	i_d (μA)	$E'_{1/2}$ (−V vs. S.C.E)	$F_{00}[(X)]$	$F_{10}[(X)] \times 10^3$	$F_{20}[(X)] \times 10^6$	$F_{30}[(X)] \times 10^9$
0.000	8.40	0.650	–	–	–	–
0.001	8.00	0.691	23.9	22.9	13.90	0.900
0.002	7.90	0.704	64.99	31.998	14.99	0.995
0.003	7.77	0.717	168.49	55.83	15.610	0.870
0.004	7.50	0.724	301.04	75.01	16.5025	0.875
0.005	7.30	0.730	494.77	98.755	17.951	0.990
0.006	7.06	0.735	723.88	120.48	18.58	0.930
0.007	6.70	0.738	1015.00	145.35	19.485	0.926

log A = 0.6020, log B = 3.95, log C = 7.1139, log D = 8.9542, C_x = concentration of Serine in moles $litre^{-1}$

Binary complexes are less stable than the ternary complexes, this is evidenced by the values of mixed-ligands complexes and stability constants.

5 Conclusions

In this paper, the interaction of Cd(II) ions and DL-Alanine, 4,4,4-trifluoro-1-(2-napthyl)butane-1,3-dione, and L-Serine was investigated using polarographic method. The stability constants values for "mixed-ligands complexes" are greater than the stability constants values for simple "metal-ligand system".

All the above ligands act as bidentate ligand. L-serine forms weaker mixed-ligands complex with Maybridge of Cd(II) than DL-Alanine due to presence of –OH group on C_3 carbon which reduces the electron density on the negatively charged oxygen atom of –COOH group.

It is seen from the results that the order of stability constant of complexes of Cd (II)-4,4,4-trifluoro-1-(napthyl)butane-1,3-dione with amino acids are:

6 DL-Alanine > L-Serine

The schaap and McMaster method was applied (Tables 5, 6 and 7) and from which the value of A, B, C, D were obtained for both sets (Table 8).

References

1. Underwood EJ (1991) Trace element in human and animal nutrition. Academic press, London
2. Weinstein B (1994) Chemistry and biochemistry of amino acids. Peptides and proteins, vol 3, no 16. Marcel Dekker, New York
3. Ottaway JH, Apps DK (1984) Biochemistry, 4th edn. W.B. Saunders Company, London
4. Nomenclature and symbolism for amino acids and peptides (1984) Pure Appl Chem 56(5): 595–624
5. David LN, Cox MM (2005) Principle of biochemistry, 4th edn. New York, pp 684–685
6. Buxbaum G, Pfoff G (2005) Industrial inorganic pigments, 3rd edn. Willey-VCH, pp 121–123
7. Smith CJE, Higgs MS, Baldwin KR (1999) Advances to protective coating and their application to ageing aircraft. Mechanical Sciences Sector, UK
8. Lohiya RK, Pratihar PL, Singh RV, Mukherjee (2001) Electrochemical studies at dme of Copper-2-Amino-Lepidine complexes in Aqueous 1, 4-Dioxan, Dimethyl Formamide, Acetonitrile and Formamide mixture. Orient J Chem 17(3):391–398
9. Verma MK, Chandel CPS (2003) Mixed ligand complexes of Cd with L-citrulline and some bicarboxylic acids: a electrochemical study. Orient J Chem 19:49–54
10. Singh SK, Chandel CPS (2003) Studies on mixed-ligand complexes of lead ion with some amino acids and phthalate at DME: a polarographic approach. Bull Electrochem 19(3):119–26
11. Jadwiga U, Henryk K, Barbora K (1992) Polarographic and potentiometric study of ternary complexes of Cadmium with Penicillamine and Histidine. J Coord Chem 25(2):149–154
12. Verma A, Chauhan PKS, Paliwal RK (2004) Polarographic study of stability constants of Cadmium(II) complexes with pyridine and some amino acids (Glycine and Alanine). Orient J Chem 20(2)
13. Verma A, Chauhan PKS, Paliwal RK (2004) Polarographic study of stability constants of Lead(II) complexes with B-Picoline and some amino acids (Lysine and Metheonine). Orient J Chem 20(2)
14. Verma M, Chandel C (2005) Stability constants of complexes of Cd(Ii) with some Bicarboxylic acids and Ascorbic acid (vitamin C): an electro analytical study. Orient J Chem 21(1):9
15. Kalpana S, Kaur T, Gupta KJ (2005) Study of complexes of cadmium(II) and lead(II) with mixed ligand (thioglycolate-oxalate). Indian Chem Soc 82:61
16. Jain AK, Khan F (1998) Polarographic study of mixed-ligand complexes of cadmium(II) with L-amino acid and vitamin B5. J Indian Chem Soc 75(1):31–32

17. Yadav RP (2003) Ph.D. thesis, University of Rajasthan, Jaipur, India
18. Khan F, Sahu PL (2000) Thermodynamics parameter of [Cd-L-Amino Acidate-Vitamine B ~ 5] system: polarographic approach. Ultra Sci Physic Soc 12(1):106–108
19. Khatoon Z, Kabir-ud-Din (1990) Potentiometric studies on mixed-ligand complexes of copper (II) and zinc(II) with some amino acids (glycine, DL-α-alanine and DL-valine) as primary ligands and imidazole as a secondary ligand. J Indian Chem Soc 15(3):217–221
20. Arishy NM, Ammar RA, Al-Warthan A (2014) Determination of stability constants of mixed ligand complexes of Picolinic acid and other bioactive ligands with Zn(II) by potentiometric titration method. Asian J Chem 26(8):2395
21. Verma A, Paliwal RK (2005) Polarographic study of stability constants of Copper(II) complexes with [beta]-picoline, Malic and Tartaric acid. Asian J Chem17(1):355
22. Verma A, Paliwal RK (2005) Polaigraphically determination of stability constants of mixed ligand complexes of Cadmium(II) with Pyridine and some Amino Acids. Asian J Chem 17(3): 1423
23. Chauhan PKS, Verma A, Paliwal RK (2005) Polarographic study of stability constants of Cadmium(II) complexes with β-picoline and some Hydroxy acids. Asian J Chem 17(1): 349–354
24. Kumar A, Chauhan PKS, Paliwal RK (2004) Stability constant of Cadmium(II) complexes with ethylenediamine and carboxylic acids: a polyrographic study. Int J Chem Sci 2(3)
25. Sharma G, Chandel CPS (2002) Electrochemical studies on mixed ligand complexes of Cadmium Ion with some Dicarboxylic acids and Pyridoxine (vitamin B6). Asian J Chem 14(1):23–33
26. Prakash D, Shafayat M, Jamal A, Gupta AK, Prakash D (2005) Mixed ligand complexes of copper glicinate with N-oxide. Orient J Chem 21(3):555–558
27. Joshi JD, Brahm Bhatt MP, Sharma S, Vora JJ (2003) Potentiometric studies and mechanism of protonation in complex formation of some Cu(II) mixed ligand complexes. Asian J Chem 15:373–378

MnO_2 Nanoparticles Embedded Polypyrrole Nanotubes for Supercapacitor Electrodes

Taruna Singh, Raj Kishore Sharma and Gurmeet Singh

Abstract This research article refers to the synthesis and characterization of MnO_2 nanoparticles embedded on polypyrrole nanotubes. The product was synthesised by chemical oxidative polymerization method. Combination of MnO_2 nanoparticles and polypyrrole nanotubes enhance the capability of the nanocomposite. The microstructures and properties of Polypyrrole nanotubes (PPy) and MnO_2 embedded Polypyrrole nanotubes (PPy:MnO_2) were determined by TEM and SEM. FT-IR spectra was recorded to determine the chemical structure of the products. XRD was used to determine the crystalline structure of the products. The PPy:MnO_2 nanocomposite electrode shows substantial improvement in the redox performance compared to individual component PPy. The specific capacitance value of electrode material PPy:MnO_2 was found to be ~ 200 F g^{-1} at 5 mV s^{-1}.

Keywords Nanoparticles · Nanotubes · Chemical oxidative polymerization Specific capacitance · Supercapacitors

1 Introduction

Supercapacitors which are also known as ultracapacitors are attracting much attention due to their longer cycling life, higher power density and higher energy density. They supply around 100 times more energy than electrolytic capacitors and also link the gap between rechargeable batteries and electrolytic capacitors. In supercapacitors conventional solid dielectric of ordinary capacitors is not used [1–3]. Supercapacitors are of two types based on the electrode material used: Electric double-layer capacitor (EDLC) and pseudocapacitors [4–6]. Supercapacitors are used in defence, space

T. Singh · R. K. Sharma (✉) · G. Singh
Department of Chemistry, University of Delhi, New Delhi 110007, Delhi, India
e-mail: drrajksharma@yahoo.co.in; rajkshama@chemistry.du.ac.in

V. S. Parmar et al. (eds.), *Green Chemistry in Environmental Sustainability and Chemical Education*, https://doi.org/10.1007/978-981-10-8390-7_19

research, consumer goods and transport vehicles like trains, planes, cars, buses, cranes and elevators which requires various rapid charge and discharge cycles. In static random-access memory (SRAM) smaller units of supercapacitors are used for memory backup.

Conducting polymers e.g. polyprrole, polyacetylene and polyaniline are promptly gaining attraction in processable materials with superior electrical and physical properties and lesser cost. The nanostructured forms of conducting polymers that is nanotubes and nanowires have higher surface area and enhanced dispersibility which lead to their use in batteries, light emitting diodes and chemical sensors [7].

Manganese oxide (MnO_2) under ambient conditions shows virtuous physical and chemical properties [8–11]. As MnO_2 shows polymorphism behaviour and flexibility in its structure, this lead to its use in various applications. Primarily MnO_2 is used as a component of zinc carbon batteries and alkaline batteries. It is also used in glass making, ceramics, biosensors, catalysis etc. [12, 13]. It is used as an excellent material for replacement of RuO_2 in supercapacitors as it is economical and also shows high specific capacitance [14–17].

We have synthesised polypyrrole nanotubes (PPy) and MnO_2 embedded on polypyrrole nanotubes (PPy:MnO_2) by chemical oxidative polymerization method. According to electrochemical studies the PPy:MnO_2 nanocomposite has superior electrochemical performance because of adsorption of MnO_2 nanoparticles on polypyrrole nanotubes. Here MnO_2 nanoparticles and polypyrrole nanotubes were used because of their eminent properties [18–20].

2 Experimental

2.1 Materials

Pyrrole (Spectrochem), Ferric Chloride (Molychem), Manganese acetate (SRL), Potassium Permanganate, Methyl Orange, Isopropoyl Alcohol and nafion binder. The chemicals which we used were of analytical grade.

2.2 Preparation of PPy and PPy:MnO_2

PPy was synthesised by in situ chemical oxidative polymerization method. 0.486 g ferric chloride was dissolved in 60 ml methyl orange (5 mM) and the solution was stirred for half an hour, a flocculent precipitate was appeared. 210 μL pyrrole monomer was added to it and the solution was stirred for 5 h at room temperature. Dark brown colour solution with black precipitate was appeared. 200 mL deionized water was added to it and the solution was kept stagnant around half an hour for termination of reaction. Polypyrrole precipitate formed was filtered using 0.2 μm PTFE membrane and the filtered precipitate was washed numerous times with ethanol/deionized water until the filtrate was colourless. The formed precipitate was then air dried in vacuum oven at 90 °C for 6 h.

PPy:MnO_2 nanocomposite was synthesised by using 135 mg of above synthesised polypyrrole nanotubes. 4 mL of manganese acetate (1.5 M) was added to it and the solution was stirred for 3 h at room temperature. 2 mL $KMnO_4$ (1 M) was added to it dropwise while stirring and stirring was continued for 4 h. Blackish brown solution with black precipitate was observed. 10 mL deionized water was added to it and the reaction mixture was kept stagnant for half an hour for termination of reaction. The prepared precipitate was filtered using 0.2 μm PTFE membrane and the formed precipitate was washed several times with ethanol/deionized water until the filtrate was colourless. The precipitate was then air dried in vacuum oven at 90 °C for 6 h. The synthesised product was black in colour and it was in powdered form.

2.3 *Characterizations*

The morphology and particle size of PPy and PPy:MnO_2 were studied by TEM (Phillips Technai T-300) and SEM (Zeiss Ultra 55). FT-IR spectrophotometer (BX Perkin Elmer) was used to identify the chemical structure of the synthesised products. XRD (D8 Discoverer) was used to identify the crystalline structure of the products. CHI 604 D electrochemical workstation was used to record electrochemical performance. AC Impedance studies was carried out by using PARSTAT 4000.

2.4 *Electrode Preparation*

10 mg each of PPy and PPy:MnO_2 were ultrasonically mixed with 10 μL of nafion binder in IPA. PPy and PPy:MnO_2 suspensions were then sprayed onto 1 cm^2 area of the polished graphite sheet. PPy and PPy:MnO_2 coated graphite sheets were dried in oven at 90 °C for 12 h. The active material loaded over the graphite sheet was found to be ~0.5 mg/cm^2 for both PPy and PPy:MnO_2. Cyclic Voltammetry study was carried out using three electrode cell assembly. Active material loaded graphite electrodes of PPy and PPy:MnO_2 were served as working electrodes. Saturated Ag/AgCl electrode was used as a reference electrode. Platinum electrode was used as counter electrode. In all the measurements 0.5 M Na_2SO_4 was used as electrolyte.

3 Results and Discussion

Figure 1 depicts the FT-IR spectra of PPy and PPy:MnO_2. In case of PPy broad peak at 3420 cm^{-1} appears which represents O–H stretching of water molecules. The peak at 1544 cm^{-1} depicts polypyrrole ring vibrations. At 1170 cm^{-1} the peak observed is due to C-N stretching vibration and 1030 cm^{-1} the peak observed is due to N-H in plane deformation vibration. At 897 cm^{-1} the peak observed is due to C-H out of plane vibration which indicates polymerization of pyrrole.

PPy:MnO_2 shows broad band at 3392 cm^{-1} which is due to O-H stretching of water molecules. A weak band at 1555 cm^{-1} is observed which is due to polypyrrole ring vibrations. The peaks at 1030 cm^{-1} is due to N-H in plane deformation vibration. The sharp peak at 569 cm^{-1} is due to Mn–O vibration which confirms the presence of MnO_2.

XRD pattern of PPy (Fig. 2) shows characteristic peaks of amorphous polypyrrole at $2\theta = 17.53°, 19.88°, 23.20°$. XRD pattern of PPy:$MnO_2$ shows very broad diffraction peaks at 36.68° and 65° with low intensity which show that the prepared nanocomposite is amorphous in nature. In PPy:MnO_2, no characteristic peak of polypyrrole is observed which shows that polypyrrole nanotubes have been covered with MnO_2 nanoparticles.

SEM micrographs of polypyrrole nanotubes indicate approximately 0.3 µm diameter as shown by red circles (Fig. 3a). However the micrographs of the prepared nanotubes exhibits varied length scale and diameter. Figure 3b shows SEM micrographs of PPy:MnO_2. Here some spherical particles are attached on the surface of polypyrrole nanotubes which indicate that MnO_2 nanoparticles have been embedded on polypyrrole nanotubes. MnO_2 adsorption on polypyrrole nanotubes are shown by blue circles.

As seen by SEM micrographs the TEM micrograph of PPy also reveals the presence of long and uniform polypyrrole nanotubes. Hollow structure of polypyrrole nanotube is shown by red circle (Fig. 4a). Such hollow structures possess large surface area which lead to enhance charge storage properties. The selected area diffraction pattern (SAED) shows amorphous nature of polypyrrole nanotubes (Fig. 4b).

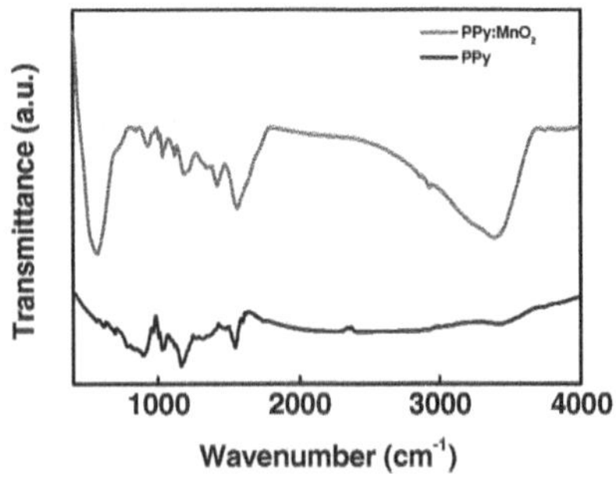

Fig. 1 FT-IR spectra of polypyrrole nanotubes (PPy) and MnO_2 embedded polypyrrole nanotubes (PPy: MnO_2)

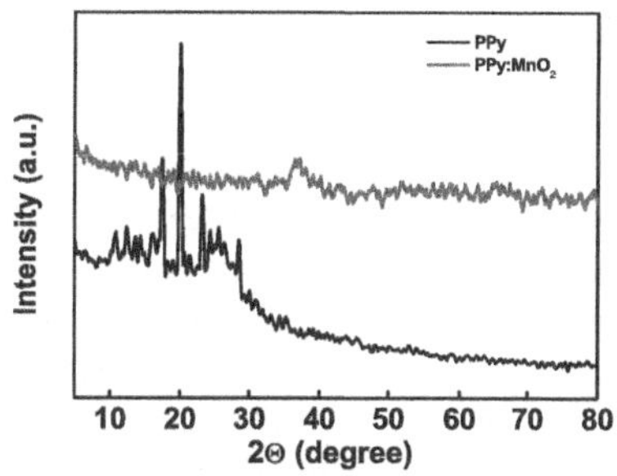

Fig. 2 XRD of polypyrrole nanotubes (PPy) and MnO_2 embedded polypyrrole nanotubes (PPy:MnO_2)

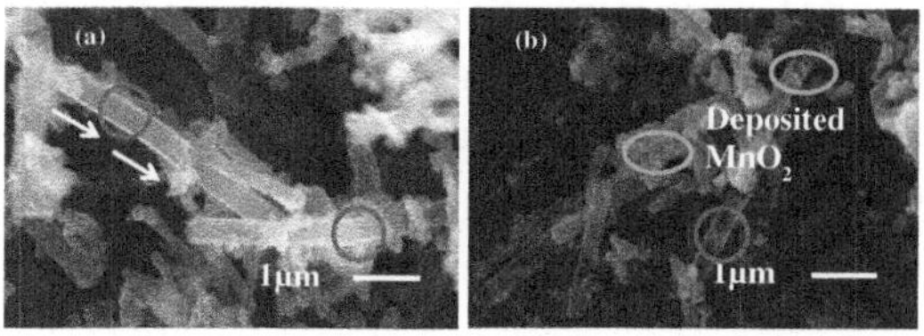

Fig. 3 SEM micrographs of Polypyrrole nanotubes (**a**) and MnO_2 embedded polypyrrole nanotubes (**b**)

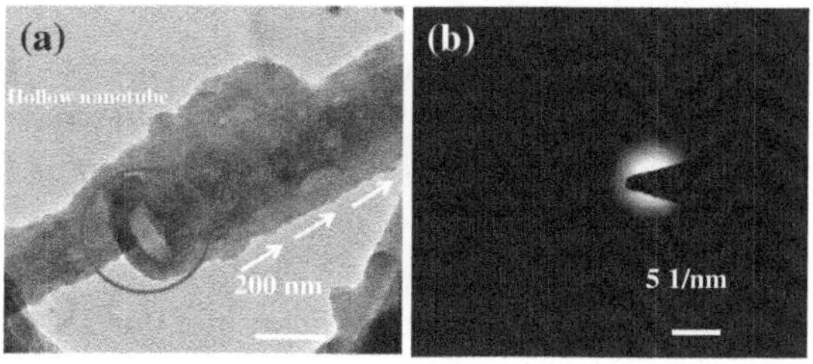

Fig. 4 TEM micrographs of polypyrrole nanotubes (PPy)

TEM micrographs of PPy:MnO_2 show that the polypyrrole nanotubes are embedded by small particles which was identified as MnO_2 nanoparticles. Here MnO_2 nanoparticles are present near the mouth of hollow polypyrrole nanotube which is shown by red circle (Fig. 5a). Due to hollow nature, the formed nanocomposite PPy:MnO_2 is showing higher specific capacitance. Attachment of MnO_2 nanoparticles on polypyrrole nanotubes results in enhanced active surface area and subsequently higher specific capacitance of the nanocomposite. Figure 5b is the selected area diffraction (SAED) pattern which shows amorphous nature of PPy:MnO_2.

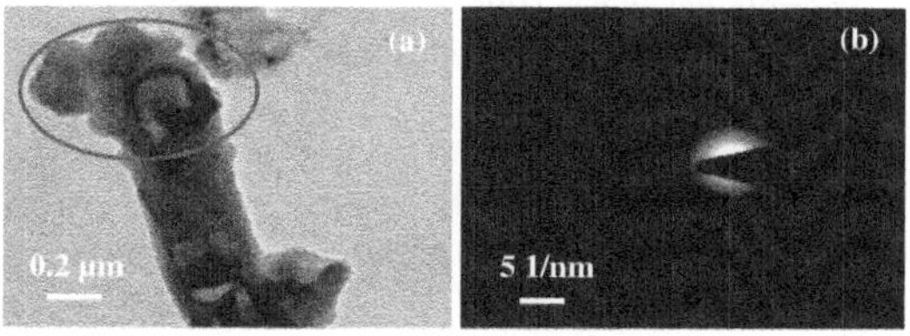

Fig. 5 TEM micrographs of MnO_2 embedded polypyrrole nanotubes (PPy:MnO_2)

4 Electrochemical Studies

The cyclic voltammograms of PPy and PPy:MnO_2 nanocomposite was measured at 5 mV s^{-1} in 0.5 M Na_2SO_4 (Fig. 6). The potential window of PPy and PPy:MnO_2 was chosen from −0.2 to 0.8 V. Specific capacitance of the products was calculated by using cyclic voltammetry from area under current-voltage curve using Eq. 1 [21]. Specific capacitance calculated from the curves was found to be ~122 and ~200 F g^{-1} for PPy and PPy:MnO_2. The specific capacitance of PPy:MnO_2 is found to be higher than that of PPy which is due to adsorption of MnO_2 nanoparticles on polypyrrole nanotubes. This adsorption plays an important role in improving the material accessibility through porous network. Polypyrrole nanotubes provide a template for the growing MnO_2 nanoparticles with large surface area while the presence of MnO_2 nanoparticles on polyyrrole nanotubes help in avoiding chain defects.

$$C_{sp} = \int_{E_1}^{E_2} i(E)dE/2(E_2 - E_1)mv \quad (1)$$

C_{sp}	Single electrode's specific capacitance in three cell assembly
E_1 and E_2	cut off potential of cyclic voltammetry
I(E)	instantaneous current, $\int_{E_1}^{E_2} i(E)dE$ = integration of positive and negative sweep in cyclic voltammograms to get total voltammetric charge
$E_2 - E_1$	width of potential window
m	mass of material deposited on electrode
v	scan rate potential

Nyquist plot depicts the plot of imaginary component (Z″) of impedance against the real component (Z′). Here Nyquist plot represents the Warburg region for PPy electrode from high to low frequency while PPy:MnO_2 electrode reveals the bend frequency (Fig. 7). Low equivalent series resistance of PPy:MnO_2 in high frequency region is accredited to balanced ionic and electronic conduction.

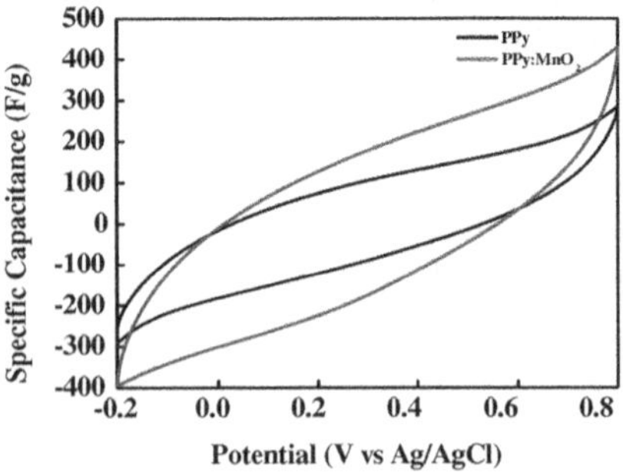

Fig. 6 Cyclic voltammograms of polypyrrole nanotubes (PPy) and MnO_2 embedded polypyrrole nanotubes (PPy: MnO_2) at 5 Mv s^{-1}

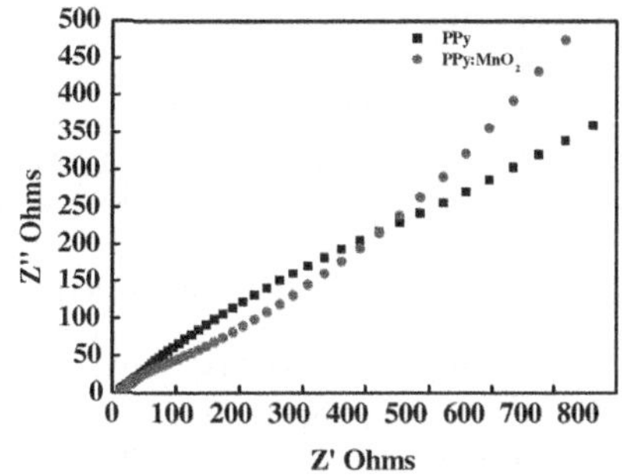

Fig. 7 Nyquist plot of polypyrrole nanotubes (PPy) and MnO_2 embedded polypyrrole nanotubes (PPy:MnO_2)

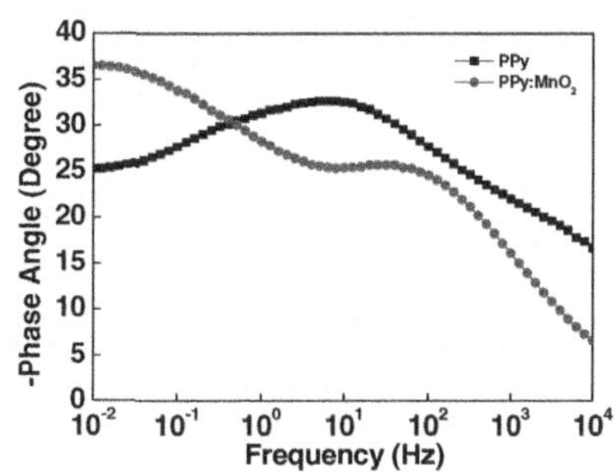

Fig. 8 Bode plot of polypyrrole nanotubes (PPy) and MnO_2 embedded polypyrrole nanotubes (PPy:MnO_2)

Figure 8 demonstrates the Bode Plot of PPy and PPy:MnO_2 in the frequency range of 0.01–1 × 10^4 Hz. Bode plot suggest phase angles at ~26.5° and ~37° which designates the worthy charge storage behaviour of PPy:MnO_2 as compared to PPy.

5 Conclusion

PPy:MnO_2 nanocomposite was synthesised by chemical oxidative polymerization method. We have explored the role of manganese oxide (MnO_2) nanoparticles dispersion on polypyrrole nanotubes toward improved performance in supercapacitor device. Electrochemical analysis shown that PPy:MnO_2 nanocomposite has superior electrochemical performance because of nanoscale dispersion of MnO_2 nanoparticles on polypyrrole nanotubes. The nano level dispersion of MnO_2 on polypyrrole nanotubes provide percolative path for electronic conduction. At 5 mV s^{-1} PPy:MnO_2 nanocomposite possess ~200 F/g specific capacitance as compared to PPy which has ~122 F/g specific capacitance. The results show that the prepared nanocomposite PPy:MnO_2 can be an effective material for supercapacitor electrodes.

Acknowledgements Authors thank all the lab members and USIC staff of Delhi University. The financial support was provided by Department of Science and Technology (DST), Delhi, India through DST Women Scientist B, Project No. DST/Disha/SoRF-PM/029/2013/G, Dated: 8/07/15.

References

1. Conway BE (1999) Electrochemical capacitors: scientific fundamental and technological applications. Kluwer, Dordrecht, The Netherlands
2. Miller JR, Simon P (2008) Electrochemical capacitors for energy management. Science 321:651–652
3. Simon P, Gogotsi Y (2008) Materials for electrochemical capacitors. Nat Mater 7:845–854
4. Kotz R, Carlen M (2000) Principals and applications of electrochemical capacitors. Electorchim Acta 45:2483–2496
5. An KH, Kim WS, Park YS, Choi YC, Lee SM, Chung DC, Bae DJ, Lim SC, Lee YH (2003) Supercapacitors using single-walled carbon nanotube electrodes. Adv Mater 13:497–500
6. Wang Q, Wen Z, Li JA (2006) Hybrid supercapacitor fabricated with a carbon nanotube cathode and a TiO_2-B nanowire anode. Adv Funct Mater 17:845–854
7. Sharma RK, Rastogi AC, Desu SB (2008) Pulse polymerized polypyrrole electrodes for high energy density electrochemical supercapacitors. Electrochem Commun 10:268–272
8. Sharma RK, Oh HS, Shul YG, Kim HS (2007) Carbon supported MnO_2 nanorods for electrochemical supercapacitor. J Power Sources Short Commun 173:1024–1028
9. Zordan TA, Hepler LG (1968) Thermochemistry and oxidation potentials of manganese and its compounds. Chem Rev 68:737–745
10. Zhang H, Cao G, Wang Z, Yang Y, Shi Z, Gu Z (2008) Growth of manganese oxide nanoflowers on vertically-aligned carbon nanotube arrays for high-rate electrochemical capacitive energy storage. Nano Lett 8:2664–2668
11. Cheng F, Zhao J, Song W, Li C, Ma H, Chen J, Shen P (2006) Facile controlled synthesis of MnO_2 nanostructures of novel shapes and their application in batteries. Inorg Chem 45:2038–2044
12. Subramanian V, Zhu H, Wei B (2008) Alcohol-assisted room temperature synthesis of different nanostructured manganese oxides and their pseudocapacitance properties in neutral electrolyte. Chem Phys Lett 453:242–249
13. Yang X, Tang W, Feng Q, Ooi K (2003) Single crystal growth of Birnessite and Hollandite-type manganese oxides by a flux method. Cryst Growth Des 3:409–415
14. Zheng D, Yin Z, Zhang W, Tan X, Sun S (2006) Novel branched—MnOOH and MnO_2 multipod nanostructures. Cryst Growth Des 6:1733–1735
15. Xu M, Kong L, Zhou W, Li H (2007) Hydrothermal synthesis and pseudocapacitance properties of MnO_2 hollow spheres and hollow Urchins. J Phys Chem C 111:19141–19147
16. Devaraj S, Munichandraiah N (2008) Effect of crystallographic structure of MnO_2 on its electrochemical capacitance properties. J Phys Chem C 112:4406–4417
17. Toupin M, Brousse T, Belanger D (2002) Influence of microstructure on the charge storage properties of chemically synthesised manganese dioxide. Chem Mater 14:3946–3952
18. Sharma RK, Rastogi AC, Desu SB (2008) Manganese oxide embedded polypyrrole nanocomposite for electrochemical supercapacitor. Electrochim Acta 53:7690–7695
19. Sharma RK, Zhai L (2009) Multiwall carbon nanotube supported poly (3,4-ethylenedioxythiophene)/manganese oxide nanocomposite electrode for supercapacitors. Electrochim Acta 54:7148–7155
20. Grover S, Shekhar S, Sharma RK, Singh G (2014) Multiwalled carbon nanotube supported polypyrrole manganese oxide composite supercapacitor electrode: role of manganese oxide dispersion in performance evolution. Electrochim Acta 116:137–145
21. Sahu V, Shekhar S, Sharma RK, Singh G (2015) Ultrahigh performance supercapacitor from lacey reduced graphene oxide nanoribbons. Appl Mater Interfaces 7:3110–3116

Remediation of Dioxins Contaminated Soil Using UV Radiations in Conjunction with Groundnut Oil

Vaishali V. Shahare and Neeta Thacker

Abstract The effect of UV radiations and oil mediated UV radiations in removal of PCDDs and PCDFs congeners was studied. The effectiveness of photolytic degradation using groundnut oil was examined for removing tetra-, and octachlorinated congeners of dioxins and furans from contaminated soil by varying the intensity of UV radiation, exposure period and analyte concentration. All the four congeners tested were successfully degraded. The maximum removal of 50.61–62.3% of 2, 3, 7, 8-TCDD, 44.1–69.24% of 2, 3, 7, 8-TCDF, 35.0–56.03% of OCDD and 37.11–59.64% of OCDF was observed at an initial concentration of 2.5 and 5.0 ng g^{-1} respectively after an UV exposure period of 24 h. The dioxin congeners in the soil sample showed decrease in concentration with the increase in the exposure time and intensity of UV radiations. More degradation of dioxin was found when the samples were exposed for 34 h using 400 W of UV lamp. The rate of photocatalytic degradation of polychlorinated dioxins was found to decrease with the increase in number of chlorines.

Keywords Dioxins · Soil contamination · Degradation · Groundnut oil

1 Introduction

Polychlorinated dioxins (PCDDs) and Polychlorinated furans (PCDFs) results on account of burning of municipal solid wastes, industrial waste, biomedical wastes, open burning of wood and building materials, and industrial production processes [1]. The use of hazardous wastes and chlorine based chemicals and pesticides containing dioxins as contaminants as well as their disposal leads to the environ-

V. V. Shahare (✉)
Rajdhani College, University of Delhi, New Delhi 110015, India
e-mail: vaishalis.du@gmail.com

N. Thacker
National Environmental Engineering Research Institute, Nagpur 440020, India
e-mail: neetathacker9@gmail.com

V. S. Parmar et al. (eds.), *Green Chemistry in Environmental Sustainability and Chemical Education*, https://doi.org/10.1007/978-981-10-8390-7_20

mental contamination. 2, 3, 7, 8-TCDD was detected in effluent samples of pulp and paper mills using chlorine for the bleaching process. In paper mills where chlorine dioxide was used as a bleaching agent, 2, 3, 7, 8-TCDD congener ranged from below the detection limit 0.05 to 0.12 ng L^{-1}/ng g^{-1} in the effluent and pulp samples [2]. There are various techniques reported for the breakdown or reduction of dioxins in environmental or disposal samples [3–5]. To reduce PCDDs and PCDFs in the source and environment, some treatment methods like advanced oxidation and biological processes have been investigated [6, 7]. The reductive dechlorination and hydroxidation of PCDDs in solvents has been reported in liquid-phase photolysis [8], and in the photolysis on solid surface, degradation rates of PCDDs and PCDFs were found to be small as compared to the liquid-phase photolysis [9–11].

In the present paper the authors have investigated an opportunity of the degradation of tetra and octachlorinated dioxins and furans by using oil mediated UV radiations with subsequent extraction of respective PCDDs and PCDFs from soils.

2 Result

The effect of varying UV radiations on the tetra-(2, 3, 7, 8) and octachorinated dioxins and furans congeners studied. The removal increased with the increase in exposure period to UV radiations. The removal of 15.3, 17.25, 11.22 and 13.25% was observed for 2, 3, 7, 8-TCDD, 2, 3, 7, 8-TCDF, OCDD and OCDF at an initial concentration of 2.5 ng g^{-1} respectively after an exposure period of 24 h using 125 W UV lamp (Fig. 3). The removal ranging from 33 to 50% was observed for 2, 3, 7, 8-TCDD, 2, 3, 7, 8-TCDF, OCDD OCDF in different set of experiments at initial concentration of 2.5 ng g^{-1} in 24 h exposure period using 400 W UV lamp (Fig. 1). The removal ranged from 24 to 32.3% for similar set of experiments conducted an initial concentration of 5.0 ng g^{-1} with 24 h exposure period using 400 W UV lamp (Figs. 1, 2 and 3).

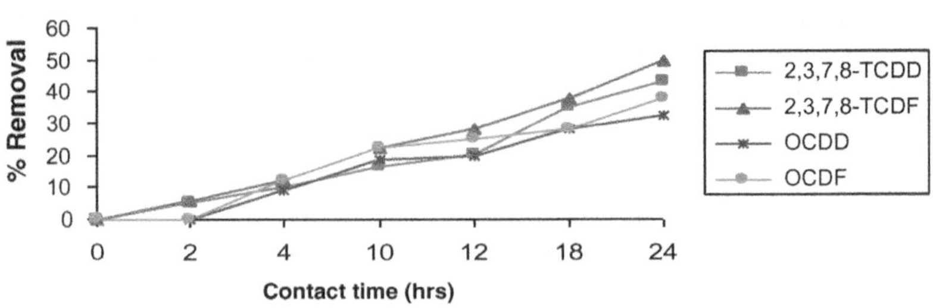

Fig. 1 Effect of UV-radiations (400 W) on removal of 2, 3, 7, 8-TCDD/F OCDD/F in soil at 2.5 ng g^{-1}

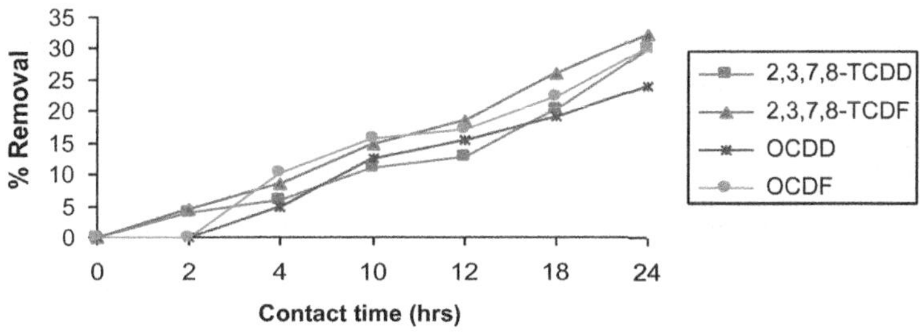

Fig. 2 Effect of UV radiations (400 W) on removal of 2, 3, 7, 8-TCDD/F OCDD/F in soil at 5.0 ng g^{-1}

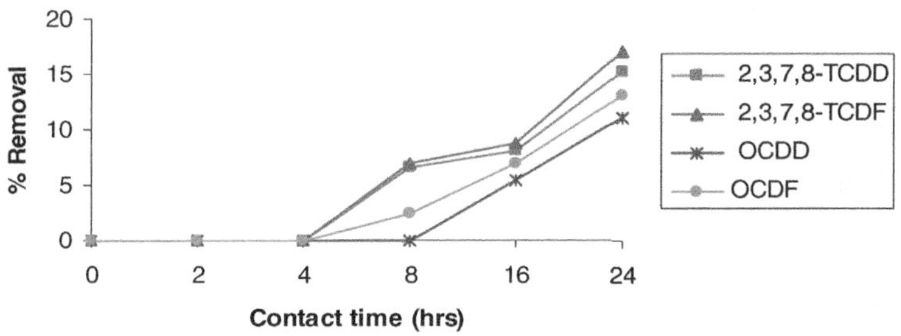

Fig. 3 Effect of UV radiations (125 W) on removal of 2, 3, 7, 8-TCDD/F OCDD/F in soil at 2.5 ng g^{-1}

2.1 Degradation

2.1.1 UV Radiations (400 W) with Oil Mediation

The degradation of all the tetra-(2, 3, 7, 8) and octachlorinated dioxins and furans were studied by varying the exposure period using groundnut oil mediated UV radiations. The degradation of 2, 3, 7, 8-TCDD increased after the addition of the oil and increase in exposure period to UV radiations. The removal of 2, 3, 7, 8-TCDD increased from 13.44 to 62.3% with increase in exposure period from 2 to 24 h (Fig. 4). However, at an initial concentration of 5.0 ng g^{-1} of 2, 3, 7, 8-TCDD, 50.61% removal was observed after an exposure period of 24 h (Fig. 5).

The maximum removal of 69.24, 56.03 and 59.64% was observed for 2, 3, 7, 8-TCDF, OCDD and OCDF respectively at an initial concentration of 2.5 ng g^{-1} after an exposure period of 24 h (Fig. 4). The maximum removal of 44.1, 35.0 and 37.11% was observed for 2, 3, 7, 8-TCDF, OCDD and OCDF at an initial concentration of 5.0 ng g^{-1} after an exposure period of 24 h (Fig. 5).

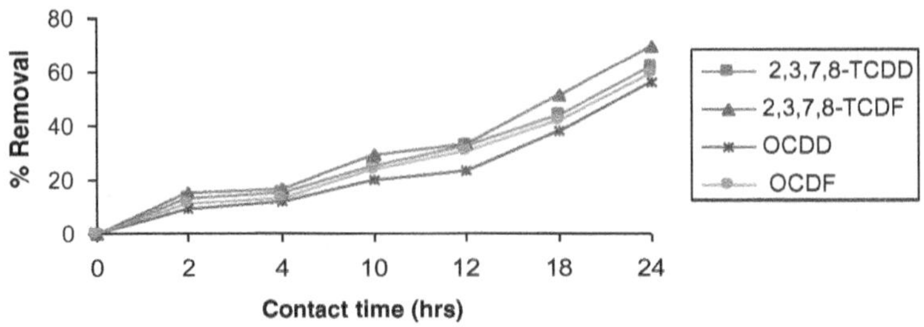

Fig. 4 Effect of UV radiations (400 W) and oil on removal of 2, 3, 7, 8-TCDD/F OCDD/F in soil at 2.5 ng g^{-1}

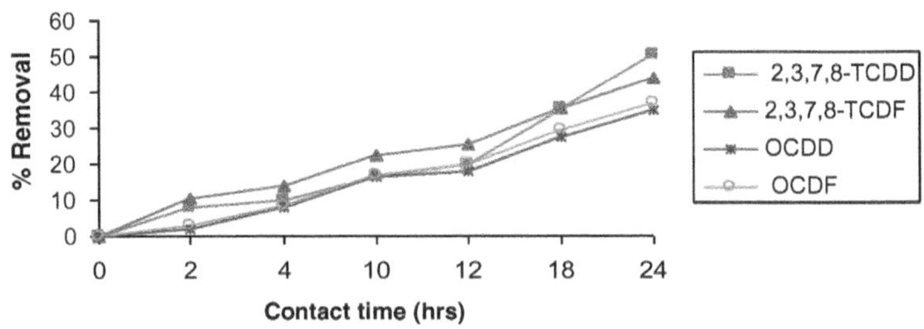

Fig. 5 Effect of UV-radiations (400 W) and oil on removal of 2, 3, 7, 8-TCDD/F OCDD/F in soil at 5.0 ng g^{-1}

2.1.2 UV Radiations (125 W) with Oil Mediation

The degradation of all the four tetra- 2, 3, 7, 8 and octachlorinated dioxins and furans increased after the oil mediation and increase in UV exposure period. However the removal was found less with decrease in UV intensity. The removal of 15, 17, 11 and 13% was observed for 2, 3, 7, 8-TCDD, 2, 3, 7, 8-TCDF, OCDD and OCDF at an initial concentration of 2.5 ng g^{-1} respectively after an exposure period of 24 h.

2.1.3 Solar Radiations

The degradation of all the four tetra (2, 3, 7, 8) and octachlorinated dioxins and furans increased with the increase in exposure period to solar radiations. The removal of 2, 3, 7, 8-TCDD increased from 5.08 to 22.36% with increase in exposure period from 8 to 34 h (Fig. 6). The maximum removal of 22.36, 32.2, 17.52 and 18.55% was observed respectively at an initial concentration of 2.5 ng g^{-1} after an exposure period of 34 h (Fig. 6).

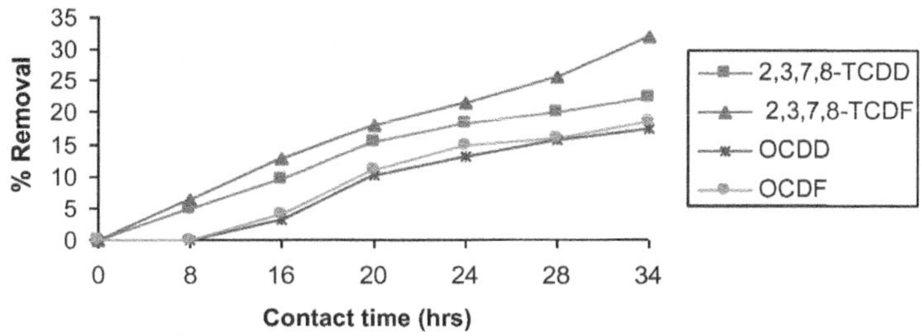

Fig. 6 Effect of solar radiations on removal of 2, 3, 7, 8-TCDD/F and OCDD/F in soil at 2.5 ng g-1

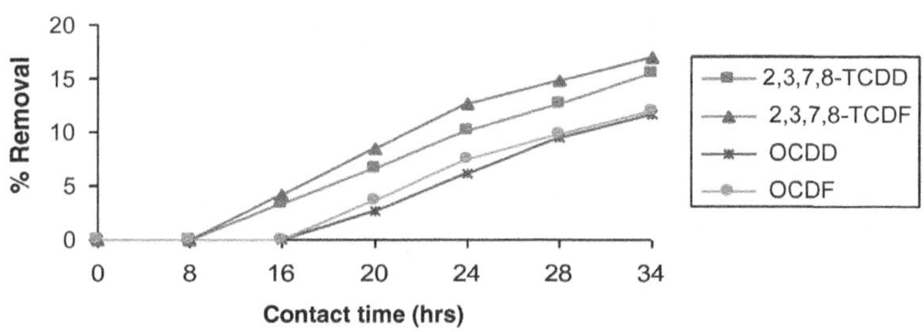

Fig. 7 Effect of solar radiations on removal of 2, 3, 7, 8-TCDD/F OCDD/F in soil at 5.0 ng g^{-1}

The removal of 15.58, 17, 11.65 and 12% was observed at an initial concentration of 5.0 ng g^{-1} of 2, 3, 7, 8-TCDF after an exposure period of 34 h (Fig. 7).

2.1.4 Solar Radiations with Oil Mediation

The degradation of all the four tetra (2, 3, 7, 8) and octachlorinated dioxins and furans increased on exposure to solar radiations with oil mediation. The removal of 2, 3, 7, 8-TCDD increased from 6.53 to 38.3% with increase in exposure period from 8 to 34 h. The maximum removal of 38.3, 49.72, 27 and 30.25% was observed for 2, 3, 7, 8-TCDD, 2, 3, 7, 8-TCDF, OCDD and OCDF respectively at an initial concentration of 2.5 ng g^{-1} after an exposure period of 34 h (Fig. 8).

The removal of 22.75 and 26.0, 17.65 and 18% was observed for 2, 3, 7, 8-TCDD, 2, 3, 7, 8-TCDF, OCDD and OCDF respectively at an initial concentration of 5.0 ng g^{-1} after an exposure period of 34 h (Fig. 9).

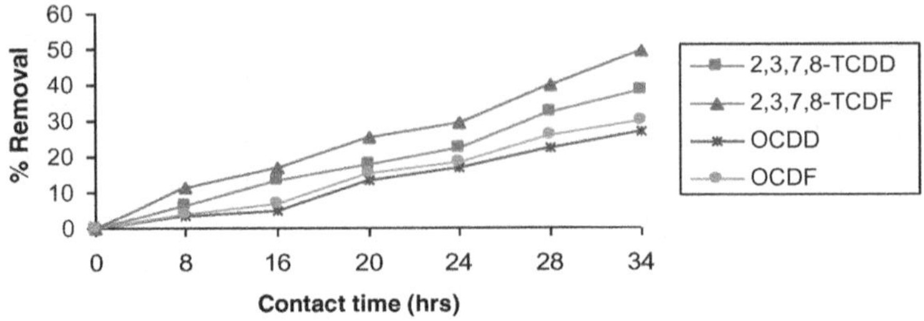

Fig. 8 Effect of solar radiations and oil on removal of 2, 3, 7, 8-TCDD/F OCDD/F in soil at 2.5 ng g^{-1}

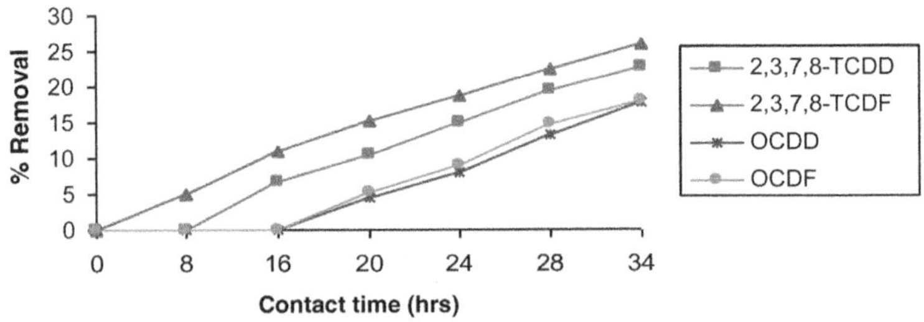

Fig. 9 Effect of solar radiations and oil on removal of 2, 3, 7, 8-TCDD/F OCDD/F in soil at 5.0 ng g^{-1}

3 Discussion

All the four congeners 2, 3, 7, 8-TCDD, 2, 3, 7, 8-TCDF, OCDD and OCDF tested were successfully degraded. The degradation was found to increase after the addition of the oil with increase in exposure period to UV radiations. However the removal was found less with decrease in UV intensity. The degradation of all the four tetra (2, 3, 7, 8) and octachlorinated dioxins and furans increased on exposure to solar radiations with oil mediation. More degradation of dioxin was found when the samples were exposed for 34 h using 400 W of UV lamp. The photocatalytic degradation rate of PCDDs was found to decrease with the increase in number of chlorines.

4 Materials and Methods

4.1 Reagents

All the chemicals used were reagent grade quality and obtained from Merck. Native and $^{13}C_{12}$-internal standard mixtures of PCDD and PCDF isomers of tetra- and octa-chlorinated-dibenzo-dioxins and furans were purchased from Cambridge Isotope Laboratories (Woburn, MA).

4.2 Experimental

Batch studies were conducted using simulated soil samples of 2, 3, 7, 8-tetrachlorodibenzo-p-dioxin (2, 3, 7, 8-TCDD), 2, 3, 7, 8-Tetrachlorodibenzofuran (2, 3, 7, 8-TCDF), Octachlorodibenzodioxin (OCDD) and Octachlorodibenzofuran (OCDF). Photochemical decomposition was carried out for a fixed period and at ambient temperature under controlled conditions using SAIC model quartz photochemical assembly with MPMV UV lamps of 125 and 400 W.

Sample preparation

The soil used for the removal experiments was grounded, sieved and dried. Soil samples (10 g) were completely dried and the dry weight was determined from the weight loss. Soil samples were spiked by known concentration of 2, 3, 7, 8-TCDD, 2, 3, 7, 8-TCDF, OCDD and OCDF and then homogenised with a stainless steel spatula. The samples were mixed and allowed to equilibrate at room temperature.

Sample analysis

The analysis of PCDDs/PCDFs (2, 3, 7, 8-TCDD, 2, 3, 7, 8-TCDF, OCDD and OCDF) in the soil samples were carried following the USEPA 1613 method [12].

4.3 Photochemical Reactor Set-Up

UV lamp

Test Soil sample of various sizes were taken in the photochemical reaction dish and exposed to the UV irradiations varying exposure time and UV intensity. The samples were analysed and the effect of photochemical reactions on the removal for 2, 3, 7, 8-TCDD, 2, 3, 7, 8-TCDF, OCDD and OCDF were established.

Sunlight

Photochemical reactor was set-up to carry out experiment using sunlight-oil mediation, Batch studies were conducted using simulated soil samples of 2, 3, 7, 8-TCDD, 2, 3, 7, 8-TCDF, OCDD and OCDF. The effect of solar radiations varying exposure time and at ambient temperature were studied by samples analyses for the residues of 2, 3, 7, 8- TCDD, 2, 3, 7, 8- TCDF, OCDD and OCDF.

The effect of UV lamp and solar irradiations on the removal of 2, 3, 7, 8-TCDD, 2, 3, 7, 8-TCDF, OCDD and OCDF with groundnut oil mediation in soil samples was studied. Control samples were also run parallel along with experimental set-up of 2, 3, 7, 8-TCDD where the samples were not exposed to UV radiations and solar radiations.

Effect of exposure period

Simulated soil samples of 2, 3, 7, 8-TCDD, 2, 3, 7, 8-TCDF, OCDD and OCDF at initial concentration of 2.5 and 5.0 ng g^{-1} were taken. 10 g Aliquotes of the samples were spread evenly on a glass plate 15 × 12 cm to form soil layer of 2 mm. The samples were placed under the UV photochemical assembly. The samples were exposed to UV lamp of 400 W for varying exposure period ranging from 0 to 34 h. The experiment was repeated using 125 W UV lamp at an initial concentration of 2.5 ng g^{-1}. The removal of each dioxin congener was calculated and the optimum exposure period was established.

The soil samples were also exposed to solar radiations for varying contact time of 0 to 34 h.

Effect of analyte concentration

Simulated soil samples, of 2, 3, 7, 8-TCDD, 2, 3, 7, 8-TCDF, OCDD and OCDF of concentration varying from 0.025 to 0.5 ng g^{-1} were prepared. The samples were exposed to UV lamp of 125, 400 W and solar radiations for fixed exposure period of 34 h.

Effect of groundnut oil and varying exposure period

Simulated soil samples, of 2, 3, 7, 8-TCDD, 2, 3, 7, 8-TCDF OCDD and OCDF of initial concentration 2.5 ng g^{-1} and 5.0 ng g^{-1} were prepared, mixed thoroughly and allowed to equilibrate at room temperature. Twenty drops of groundnut oil was added to the samples and exposed to UV lamp (400 W) for varying exposure period of 0 to 34 h.

After the exposure period the samples were allowed to cool and withdrawn and stored in airtight container until analysis. The samples were analysed as per the EPA 1613 method.

5 Conclusion

The study was aimed to assess the efficiency of oil-mediated mobilization and photodegradation of PCDDs and PCDFs as a means of decontamination of soil. A series of experiments were conducted to meet this objective.

In vegetable oil, rapid photodegradation of PCDDs and PCDFs was established. In treating the contaminated soil by photodegradation, the main barrier to photodegradation was the strong binding of PCDDs and PCDFs to soil particles. This barrier was overcome by ensuring a proper solubilisation of PCDDs and PCDFs, by using a longer exposure period and periodical mixing of the soil-vegetable oil mixture.

Acknowledgements Authors are grateful to the Instrumentation Division, National Environmental Engineering Research Institute, Nagpur, India, for providing the facilities to perform the work.

Conflicts of Interest The authors declare no conflict of interest.

References

1. Hiraoka M, Okajima S (1994) Source control technologies in MSW incineration plants. Organohalogen Compd 19:275
2. Thacker NP, Nitnaware VC, Das SK, Devotta S (2007) Dioxin formation in pulp and paper mills of India. Environ Sci Pollut Res Int 14(4):225–226
3. Parsons JR, de Bruijne JA, Weiland AR (1998) Biodegradation pathway of 2-chlorodibenzo-p-dioxin and 2-chlorodibenzofuran in the biphenyl- utilising strain JB1. Chemosphere 37:1915–1922
4. Nam P, Liu Q, Kapila S et al (1999) Assessment radiolysis and chemical dehalogenation for decontamination of PCBs and PCDDs in soil. Organohalogen Compd 40:507–510
5. Nakamiya K, Ishii K, Yoshizaki K, Furuichi T (2000) Solvent-washing of dioxin-contaminated soil and ultraviolet treatment of the extracted dioxins. Organohalogen Compd 45:423–426
6. Foester CJBD, Hites RA (1992) Photodegradation of polychlorinated dioxins and dibenzofurans adsorbed to fly ash. Environ Sci Technol 26:502–507
7. Adriaens P, Fu Q, Grbic-Galic D (1995) Bioavailibility and transformation of highly chlorinated dibenzo-*p*-dioxins and dibenzofurans in anaerobic soils and sediments. Environ Sci Technol 29:2252–2260
8. Buser HR (1976) Preparation of qualitative standard mixtures of polychlorinated dibenzo-*p*-dioxins and dibenzofurans by ultraviolet irradiation of the octachloro compounds. J Chromatogr 129:303–307
9. Nestrick TJ, Lamparski LL (1980) Identification of tetrachlorodibenzo-p-dioxin isomers at the 1-ng level by photolytic degradation and pattern recognition techniques. Anal Chem 52:1865–1874

10. Choudry GG, Webster GRB (1987) Environmental photochemistry of polycholorinated dibenzofurans (PCDFs) and dibenzo-*p*-dioxins (PCDDs): a review. Toxicol Environ Chem 14:43–61
11. Buser HR (1988) Rapid photolytic decomposition of brominated and brominated/chlorinated dibenzodioxins and dibenzofurans. Chemosphere 17:889–903
12. EPA Method-1613, (1995) Tetra-through octa-chlorinated dioxins and furans by isotope dilution by HRGC/HRMS, U.S. Environmental Protection Agency, Washington, DC

Preparation, Characterization and Optimization of Cinnamon-Loaded PLGA Nanoparticles

Vinod Kumari and Aditi Sangal

Abstract Cinnamon act as potential drug for lot of curable diseases such as Diabetes. For increasing its bioavailability as a drug, Nano encapsulation of cinnamon powder was carried out using solvent evaporation method. The effect of organic solvent and surfactant such as PVA(Polyvinylalcohol) and Pluronic F-68 was studied. The morphology and particle size of nanoparticles was studied using Scanning Electron Microscope (SEM). Drug release study for cinnamon and corresponding nanoparticles was also done. Results show that different surfactants have different impact on nanoparticles size and morphology. Nanoparticles obtained were in size range of 100–250 nm. The study reveals that cinnamon loaded PLGA nanoparticles has potential therapeutic efficiency and may be used in treatment of type II Diabetes.

Keywords Bioavailability · Nanoparticles · Surfactants · Drug Cinnamon · PLGA · PVA · Pluronic-68

1 Introduction

Diabetes leads to high sugar levels in the blood. Glucose absorbed in the cells of a human body provides energy with the help of Insulin. As there are two kinds of diabetes, in type 1 body does not synthesise insulin while in type 2 diabetes, the insulin formed is not properly used by the human body. The amount of glucose keeps on increasing with time, if there is not enough insulin present in the cells, which leads to various health issues such as eye infections, kidneys failures and nerves disorders. Diabetes can also affect chronic heart patients. Foetus can also be

V. Kumari (✉) · A. Sangal
Department of Chemistry, AIAS, Amity University, Noida, India
e-mail: dahiyavinod123@gmail.com

A. Sangal
e-mail: sangaladiti@gmail.com

V. S. Parmar et al. (eds.), *Green Chemistry in Environmental Sustainability and Chemical Education*, https://doi.org/10.1007/978-981-10-8390-7_21

affected, if mother is diabetic. Cinnamon (Cinnamomumzeylanicum, and Cinnamon cassia), the tree of tropical medicine, belongs to the Lauraceae family. Cinnamaldehyde, Cinnamic acid, and Cinnamate are found in Cinnamon and their extracts work as an antioxidant, anti-inflammatory, antidiabetic, antimicrobial, anticancer, lipid-lowering, and cardiovascular-disease-lowering compound. Blood glucose as well as cholesterol level are also decreased by cinnamon extracts. As compared to other spices, cinnamon is found to contain 20 fold higher potential, so it could be used as an anti- diabetic formulation. Polyphenols present in cinnamon include rutin (90.0672%), catechin (1.9%), quercetin (0.172%), kaempferol (0.016%), andisorhamnetin (0.103%). Cao et al. [1] demonstrated that the aqueous extract of cinnamon containing polyphenols purified by high performance liquid chromatography (HPLC) showed insulin-like activity. The aqueous extract of cinnamon markedly decreased the absorption of alanine in the rat intestine. Alanine present in Cinnamon extract helps in gluconeogenesis, is converted to pyruvate in the liver, and during gluconeogenesis, it is being used as substrate. However, Pasupuleti et al. (2014), conducted a study on diabetic postmenopausal women giving them cinnamon supplements showed poor glycemic control even though cinnamon is generally believed to be useful for diabetes. However, it is plausible that differences in the dose of cinnamon used, as well as baseline glucose and lipid levels, have led to these variations [2]. However Cinnamon has good antibiotic activity but it has low bioavailability, so biodegradable nanoformulation is done which can increase its efficacy, leads to rapid onset of therapeutic action and also increase bioavailabity of drug. Nanoparticle formulation reduces the toxicity of drug and it also allows slow and sustained drug release.

2 Results

This paper focus on preparation and characterisation of cinnamon loaded nanoparticles.

2.1 *Infrared Studies*

The characteristic peak of cinnamon loaded nanoparticles was studied by IR Spectroscopy. The IR spectra of PLGA-loaded cinnamon nanoparticles are presented in Fig. 1b. The characteristic peaks of PLGA are observed at 2992, 2942, 1758, 1460, 1422, 1370, 1130, 1086, 730, 548 cm^{-1}. Besides the characteristic groups for copolymer PLGA, the spectra show all the characteristic groups of cinnamon Fig. 1a. The peak shows there is no interaction between the drug and polymer.

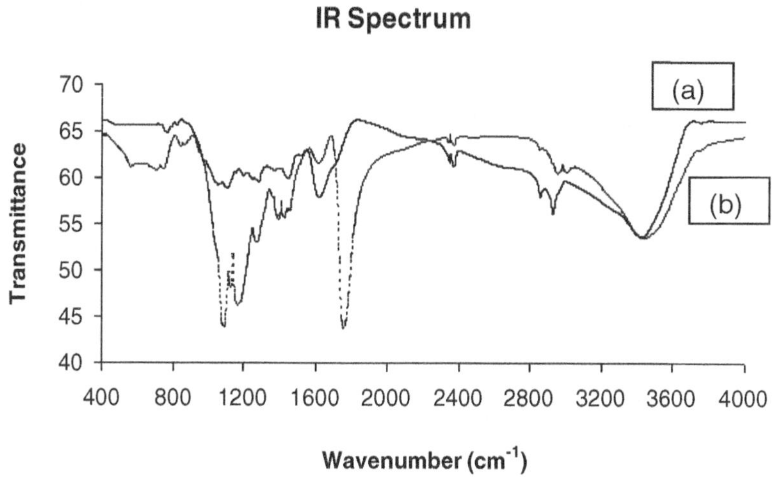

Fig. 1 IR spectrum of **a** Cinnamon **b** Cinnamon loaded PLGA nanoparticles

2.2 *Characterization of Nanoparticles Through SEM*

The morphology of cinnamon-loaded PLGA nanoparticles is shown in Fig. 2a, b. The mean particle size of the nanoparticles was between 100–250 nm when cinnamon/PLGA ratio was 1:5 and 0.1% PVA was used. SEM studies showed that distinct spherical nanoparticles are formed.

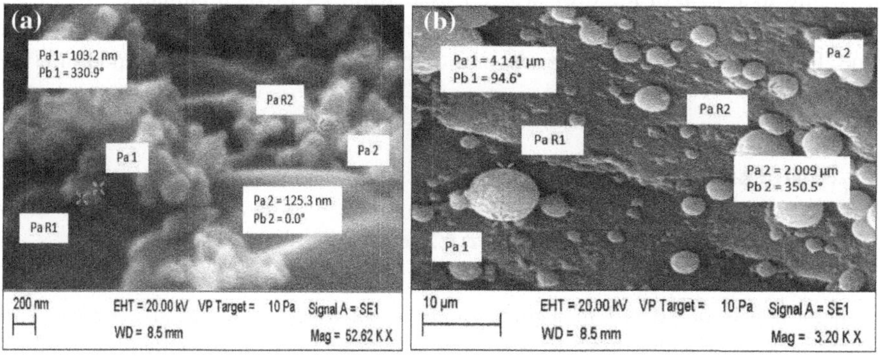

Fig. 2 SEM images of particles having different ratio of PLGA and Cinnamon **a** 1:5 **b** 1:10

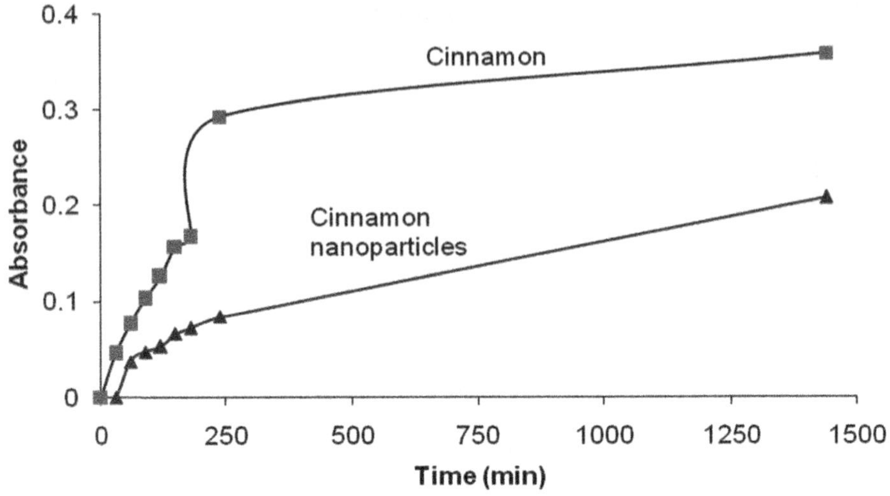

Fig. 3 Comparitive curves for dependence of absorption maxima (λ = 280 nm) for the time of release from cinnamon and cinnamon loaded PlGA nanoparticles

2.3 Invitro Release Kinetics

The release of cinnamon from PLGA-loaded cinnamon nanoparticles has been examined with UV-spectroscopy. Figure 3 shows the comparative curves for the dependence of the absorption maximum at 280 nm on the time of release of cinnamon itself as well as from cinnamon loaded-PLGA nanoparticles. It is clear from the figure that 40% of the drug was released within 4 h when cinnamon was used whereas the release was only 16% with cinnamon loaded PLGA during the same time. Even after 24 h, the drug was not completely released into the medium.

3 Materials and Method

3.1 Preparation of Cinnamon Extract

Cinnamon was powdered in a mixer and sieved to get uniform size range. This sieved powder was dissolved in methanol was used for further extraction. 8 gm of the powder was soaked in 100 ml methanol for 72 h. This preparation was then filtered through a Whatman No. 1 filter paper and the solvent was allowed to evaporate from the filtered extract at room temperature. The dried extract so obtained was stored in bottles and refrigerated for further use.

3.2 Materials

The polymer Poly (d,l-lactide-coglycolide) (PLGA), with a copolymer ratio of DL-lactide to glycolide of 50:50 (M_w = 24,000–38,000, Sigma-Aldrich) was used in this study. The surfactant used was polyvinyl alcohol (PVA) (87–90% hydrolyzed and average M_w = 30,000–70,000, Sigma-Aldrich) and Pluronic F68. The organic solvent was acetonitrile (Qualigens, Mumbai). Methanol, sodium dihydrgen phosphate, disodium hydrogen phosphate and trisodium phosphate used were of reagent grade. Nucleopore membrane was used for invitro release studies.

3.3 Preparation of Cinnamon Loaded PLGA Nanoparticles

Nanoparticles were prepared by solvent evaporation method. 10 mg of dried extract and 50 mg of PLGA were co-dissolved in 10 ml organic solvent acetonitrile. The organic phase was added drop wise to 25 ml deionised water containing 0.1% PVA using homogenizer at 50 W. The nanoparticles solution was stirred for 4 h to evaporate acetonitrile. After that suspended NP's are centrifuged (REMI, INDIA) for 20 min at 15,000 rpm, washed with deionized water and dried to obtain the dry nanoparticles.

3.4 I.R. Spectroscopy

The analysis of samples was done with IR Spectroscopy. The spectra were recorded in the range 400–4000 cm^{-1}.

3.5 Scanning Electron Microscopy

The morphology of cinnamon-loaded PLGA nanoparticles was observed using Zeiss EVOMA10 scanning electron microscope (SEM). The samples were sputter coated with a palladium layer using an ion sputter. Observations were carried out at an accelerating voltage of 20 kV and 10 Pa after 30 nm palladium coating.

3.6 Invitro Release Kinetics

Polycarbonate Track Etched membrane (Nuclepore) with pore size of 0.8 μm and 47 mm in diameter was used for invitro drug release of cinnamon nanoparticles.

The cinnamon loaded nanoparticles were placed in the glass tube, across which the nucleopore membrane was stretched. The glass tube was immersed into a beaker containing 20 ml of phosphate buffer saline (PBS) at pH 6.8. The entire system was kept at room temperature with magnetic stirring at 100 rpm. At specified time intervals (30 min, 1, 2, 4, 8, 24 h) 1 ml of solution was withdrawn and was replaced by fresh buffer. The withdrawn samples were diluted with phosphate buffer and the amount of drug dissolved was analyzed by UV-spectrophotometer at 280 nm. Drug release studies for cinnamon were also carried out and compared with the developed nanoparticles.

3.7 Ultraviolet Spectroscopy

The UV measurements were performed on double beam spectrophotometer of Systronics, India, in the range 200–400 nm.

4 Conclusions

The present study has been a satisfactory attempt to encapsulate cinnamon loaded PLGA spherical nanoparticles in the range of 100–270 nm. Different surfactants had remarkable influence on the characteristics of nanoparticles. The in vitro drug release studies show that the cinnamon loaded PLGA nanoparticles can be used for delayed and controlled drug release.

Acknowledgements Authors would like to extend their thanks to Dr. Rakesh Sharma from Delhi University for providing the IR facility and also to Dr. V. V. Ramamurthy from Indian Agricultural Research Institute, New Delhi for providing the SEM facility for their sample study.

References

1. Cao H, Polansky MM, Anderson RA (2007) Arch Biochem Biophys 459:214–222
2. Pasupuleti VR, Siew HG, Cinnamon (2014) A multifaceted medicinal plant PMC in 2014. https://doi.org/10.1155/2014/642942

Author Index

V. S. Parmar et al. (eds.), *Green Chemistry in Environmental Sustainability and Chemical Education*, https://doi.org/10.1007/978-981-10-8390-7

The manufacturer's authorised representative in the EU is Springer Nature Customer Service Centre GmbH, Europaplatz 3, 69115 Heidelberg, Germany. If you have any concerns regarding our products, please contact ProductSafety@springernature.com

Printed and bound by CPI Group (UK) Ltd, Croydon, CR0 4YY

29/07/2026

02180626-0004